HPLC

HPLC

A Practical User's Guide

Marvin C. McMaster

 **WILEY-VCH**

New York • Chichester • Weinheim • Brisbane • Singapore • Toronto

Marvin C. McMaster, Ph.D.
2070 Cordoba Drive
Florissant, MO 63033

Technical illustrations by Christopher A. McMaster

This book is printed on acid-free paper.

For ordering and customer service, call 1-800-CALL-WILEY.

Originally published as ISBN 1-56081-636-8

Library of Congress Cataloging-in-Publication Data:

McMaster, Marvin C.
    HPLC, a practical user's guide / Marvin C. McMaster.
        p.   cm.
    Includes bibliographical references and index.
    ISBN 0-471-18586-8 (acid-free)
  1. High performance liquid chromatography.   I. Title
QD79.C454M36   1994
543'.0894—dc20   93-42139
                                                       CIP

Printed in the United States of America.
10 9 8 7 6 5 4

Preface

High-pressure liquid–solid chromatography (HPLC) is rapidly becoming the method of choice for separations and analysis in many fields. Almost anything you can dissolve can be separated on some type of HPLC column. However, with this versatility comes the necessity to think about the separation desired and the best way to control it. HPLC is not now and probably will not soon be a turn-key, push-button type of operation. This is not necessarily the worst thing in the world, for it does create a great deal of job security for chromatographers.

Fortunately, controlling separations is not nearly as complicated as much of the literature may make it seem. My aim is to cut through much of the detail and theory to make this a usable technique for you. The separation models I present are those that have proved useful to me in predicting separations. I make no claims for their accuracy, except that they work. There are many excellent texts on the market, continuously updated and revised, that present the history and the current theory of chromatographic separations.

This book was written to fill a need; hopefully, your need. It was designed to help the beginning as well as the experienced chromatographer in using HPLC as a tool. Fifteen years in HPLC, first as a user, then in field sales and application support for HPLC manufacturers, has shown me that the average user wants an instrument that will solve problems, not create new ones.

I will be sharing with you my experience gained through using my own instrument, through troubleshooting customer's separations, and from field demos; the tricks of the trade. I hope they will help you do better, more rapid separations and methods development. Many of my suggestions are based on tips and ideas from friends and customers. I apologize for not giving them credit, but the list is long and my memory is short. It has been said that plagiarism is stealing from one person and research is borrowing from many. This book has been heavily researched and I would like to thank the many who have helped with that research. I hope I have returned as much as I have borrowed.

I have divided this course into three parts. The first part should give you enough information to get your system up and running. Put the book down and shoot some samples. You know enough now to use the instrument without hurting it or yourself. When you have your feet wet (not literally I hope), come back and we will take another run at the course work.

Part II shows you how to make the best use of the common columns and how to keep them up and running. (Chapter 6 on column healing should pay for the book by itself.) It discusses the various pieces of HPLC equipment, how they go together to form systems, and how to systematically troubleshoot system problems.

Finally, in Part III, we will talk about putting the system to work on real world applications. We will look at systematic methods development, both manual and automated. Then, we will look at the logic behind many of the separations that others have already made. Finally, I will discuss how to hook up to a computer and my best guess of the future of HPLC columns, systems, and detectors, including the LC/MS.

A brief note is required about the way I teach. First, I have learned that repetition is a powerful tool, not a sign of incipient senility as many people have hinted. Second, I have found in lecturing that few people can stand more than 45 minutes of technical material at one sitting. However, I have also learned that humor can sometimes act as a mental change of pace. Properly applied it allows us to continue with the work at hand. So occasionally I will tiptoe

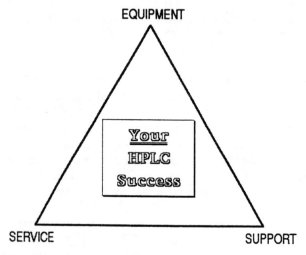

Figure P.1. Liquid chromatography success triangle.

around the lab bench. I do not apologize for it, but I thought you ought to know.

The instrument itself is the most effective teacher. Think logically about the system and the chemistry occurring inside the column. You will be surprised how well you will be able to predict and control your separation.

Remember! HPLC is a versatile, powerful, but basically simple separation tool. It is a time machine that can speed your research and, thereby, allow you to do many things not possible with slower techniques. When I finish I hope you will have the confidence to run your instrument, make your own mistakes, and be able to find your own solutions.

Your HPLC success depends on three things:

1. The suitability of the equipment you buy,
2. your ability to keep it up and running (or find someone else who will service it), and
3. the support you receive, starting out in new directions or in solving problems that come up.

Marvin C. McMaster
Florissant, Missouri
January 1994

Contents

Part I
An HPLC Primer

CHAPTER

1

Why HPLC? Advantages and Disadvantages

The first thing we need to understand is how an HPLC system works, its best applications and its advantages over other ways of separating compounds, and other techniques that might complement or even replace it. Is there a faster, easier, cheaper, or more sensitive method of achieving your results? The answer is yes, no, maybe. It really depends on what you are trying to achieve.

HPLC's virtue lies in its versatility! I have used it to separate compounds with molecular weights from 54 to 450,000. Amounts of material to be detected can vary from picograms and nanograms (analytical scale) to micrograms and milligrams (semipreparative scale) to multigrams (preparative scale). There are no requirements for volatile compounds or derivatives. Aqueous samples can be run directly after a simple filtration. Compounds with a wide polarity range can be analyzed in a single run. Thermally labile compounds can be run. I had one customer whose company made explosives for primers. Her first job each day was to explode samples of the previous day's run with a rifle. Her second job was to carry out an HPLC analysis of that day's run.

An HPLC system offers a combination of speed, reproducibility, and sensitivity. Typical runs take from 10 to 30 min, but long gradients might consume 1–2 hr. I have seen 15- to 30-sec stat runs on 3-μm columns in hospital laboratories. Retention times on the same column, run to run, should reproduce by 1%. Two columns of the same type from the same manufacturer should give 5% or better retention time reproduction on the same standard.

While the HPLC can be used in a variety of research and production operations, there are a few places where it really shines. Because it can run underivatized mixtures, it is a great tool for separating and *analyzing crude mixtures*

3

with minimum sample preparation. I began my career analyzing herbicide production runs as a method of troubleshooting product yield problems. HPLC was routinely used in the QC laboratory to fingerprint batches of final product using a similar analysis. I have helped my customers run tissue extracts, agricultural runoff waters, urine, and blood samples with minimum cleanup. These samples obviously are not very good for columns whose performance degrades rapidly under these conditions. Columns can usually be restored with vigorous washing, but an ounce of prevention is generally more effective than a pound of cure and also much more time effective.

Standards purification is another role in which HPLC excels. It is fairly easy to obtain microgram to milligram quantities of standards using the typical laboratory system.

Finally, used correctly, HPLC is a great tool for rapid *reaction monitoring* either in glassware or in large production kettles. I started my analytical career with a castoff system from the Analytical Department and a 15-min training course. I was given an existing HPLC procedure for my compound and turned loose. I was obtaining research information the next day. I could see starting material disappear, intermediates form, and both final product and byproducts appear. It was like having a window on my reaction flask through which I could observe the chemistry of the ongoing synthesis. I later used the same technique to monitor a production run in a 6000-gallon reactor. The sampling technique was different, but the HPLC analysis was identical.

Versatility, however, brings with it challenge. An HPLC is easily assembled and easily run, but to achieve optimum separation, the operator needs to understand the systems, its columns, and the chemistry of the compounds being separated. This will require a little work and a little thought, but the skills required do offer a certain job security.

I do not want to leave you with the impression that I feel that HPLC is the perfect analytical system. The basic system is rather expensive compared to other analytical tools, columns are expensive and have a relative short operating life, solvents are expensive, and disposal of used solvent is becoming a real headache. Other techniques offer more sensitivity of detection or improved separation for certain types of compounds (i.e., volatiles by GLC, large charged molecules by electrophoresis). Nothing else that I know of, however, offers the laboratory the wide range of separating modes, the combination of qualitative and quantitative separation, and the basic versatility of the HPLC system.

1.1 How Does It Work?

The HPLC separation is achieved by injecting the sample dissolved in solvent into a stream of solvent being pumped into a column packed with a solid separating material. The interaction is a liquid–solid separation. It occurs when a mixture of compounds dissolved in a solvent can either stay in the solvent or adhere to the packing material in the column. The choice is not a simple one since compounds have an affinity for both the solvent and the packing.

On a reverse-phase column, separation occurs because each compound has a different partition rate between the solvent and the packing material. Left alone each compound would reach its own equilibrium concentration in the solvent and on the solid support. However, we upset conditions by pumping fresh solvent down the column. The result is that components with the highest affinity for the column packing stick the longest and wash out last. This differential washout or elution of compounds is the basis for HPLC separation. The separated, or partially separated, discs of each component dissolved in solvent move down the column slowly moving farther apart and elute in turn from the column into the detector flow cell. These separated compounds appear in the detector as peaks that rise and fall when the detector signal is sent to a recorder or computer. This peak data can be used either to quantitate, with standard calibration, the amount of each material present, or to control the collection of purified material in a fraction collector.

1.1.1 A Separation Model of the Column

Since the real work in an HPLC system occurs in the column, it has been called the heart of the system. The typical column is a heavy-walled stainless steel tube (25-cm long with a 3–5 mm i.d.) equipped with large column compression fittings at either end (Fig. 1.1).

Immediately adjacent to the end of the column, held in place by the column fittings, is a porous, stainless steel disc filter called a *frit*. The frit serves two purposes. It keeps injection sample particulate matter above a certain size from entering the packed column bed. It also serves, at the outlet end of the column, as a bed support to keep the column material from being pumped into the tubing connecting to the detector flowcell. Each column end-fitting is drilled out to accept a zero dead volume compression fitting, which allows the column to be connected to tubing coming from the injector and going out to the detector.

The most common HPLC separation mode is based on differences in compound polarity. A good model for this polarity partition, familiar to most first-

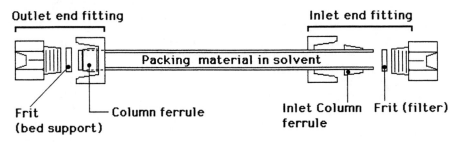

Figure 1.1.   HPLC column design.

year chemistry students, is the separation that takes place in a separatory funnel using immiscible liquids such as water and hexane. The water (very polar) has an attraction for polar compounds. The lighter hexane (very nonpolar) separates from the water and rises to the top in the separating funnel as a distinct upper layer. If you now add a purple dye made up of two components, a polar red compound and a nonpolar blue compound and stopper and shake up the contents of the funnel, a separation will be achieved (Fig. 1.2). Like attracts like.

The polar solvent attracts the more polar red compound forming a red lower layer. The blue nonpolar dye is excluded from the polar phase and dissolves in the relatively nonpolar upper hexane layer. To finish the separation we simply remove the stopper, open the separatory funnel's stopcock, draw off the aqueous layer containing the red dye, and evaporate the solvent. The blue dye can be recovered in turn by drawing off the hexane layer.

The problem with working with separatory funnels is that the separation is generally not complete. Each component has an equilibration concentration in each layer. If we were to draw off the bottom layer and dry it to recover the red dye, we would find it still contaminated with the other component, the blue dye. Repeated washings with a fresh lower layer would eventually leave only insignificant amounts of contaminating red dye in the top layer, but would also remove part of the desired blue compound. Obviously, we need a better technique to achieve our separation.

The HPLC column operates in a similar fashion. The principle of "like attracting like" still holds. In this case our nonpolar layer happens to be a moist, very fine, bonded-phase solid packing material tightly packed in the column. Polar solvent pumped through the column, our "mobile phase," serves as the second immiscible phase. If we dissolve our purple dye in the mobile phase, then inject the solution into the flow from the pump to the column, our two compounds will again partition between the two phases. The more nonpolar blue dye will have a stronger partition affinity for the stationary phase. The

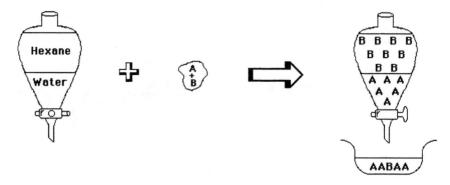

Figure 1.2.   Separation model 1 (separatory funnel).

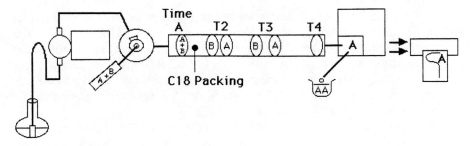

Figure 1.3. Separation model 2 (column).

more polar red dye favors the mobile phase, moves down the column more rapidly than the blue dye, and emerges first from the column into the detector. If we could see into the column we would see a purple disc moving down the column, gradually separating into a fast moving red disc followed by a slower moving blue disc (Fig. 1.3).

1.1.2 Basic Hardware: A Quick, First Look

The simplest HPLC system is made up of a high-pressure solvent pump, an injector, a column, a detector, and a data recorder (Fig. 1.4).

Note: The high pressures referred to are of the order of 2000–6000 psi. Since we are working with liquids instead of gases, high pressures do not pose an explosion hazard. Leaks occur on overpressurizing; the worst problems to be expected are drips, streams, and puddles.

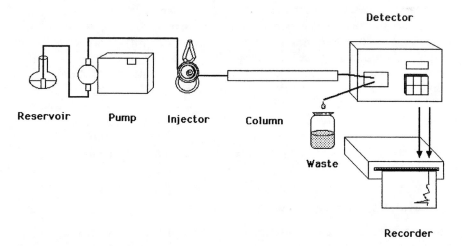

Figure 1.4. An isocratic HPLC system.

Solvent *(mobile phase)* from a solvent reservoir is pulled up the solvent inlet line into the pump head through a one-way check valve. Pressurized in the pump head, the mobile phase is driven by the pump against the column back-pressure through a second check valve into the line leading to the sample injector. The pressurized mobile phase passes through the injector and into the column where it equilibrates with the stationary phase and then exits to the detector flow cell and out to the waste collector.

The sample, dissolved in mobile phase or a similar solvent, is first loaded into the sample loop, then injected by turning a handle swinging the sample loop into the pressurized mobile phase stream. Fresh solvent pumped through the injector sample loop washes the sample onto the column head and down the column.

The separated bands in the effluent from the column pass through the column exit line into the detector flow cell. The detector reads concentration changes as changes of signal voltage. This change of voltage with time passes out to the recorder over the signal cable and is traced on paper as a *chromatogram* allowing fractions to be detected as rising and falling peaks.

There are always two outputs from a detector, one electrical and one liquid. The electrical signal is sent to the recorder for display and quantitation *(analytical mode)*. The liquid flow from the detector flow cell consists of concentration bands in the mobile phase. The liquid output from nondestructive detectors can be collected and concentrated to recover the separated materials *(preparative mode)*.

It is very important to remember that an HPLC system is both an analytical and a preparative tool. Often the preparative capabilities of the HPLC are overlooked. While normal analytical injections contain picogram to nanogram quantities, HPLCs have been used to separate as much as 1 lb in a single injection (obviously by a candidate for the *Guinness Book of World Records*). Typical preparative runs inject 1–3 g to purify standard samples.

To be effective, the detector must be capable of responding to concentration changes in all of the compounds of interest, with a sensitivity sufficient to measure the component present in the smallest concentration. There are a variety of HPLC detectors. Not all detectors will see every component separated by the column. The most commonly used detector is the UV detector, which seems to have the best combination of compound detectability and sensitivity. Generally, the more sensitive the detector, the more specific it is and the more compounds it will miss. Detectors can be used in series to gain more information while maintaining sensitivity for detection of minor components.

1.1.3 Use of Solvent Gradients

Solvent gradients are used to modify the separations achieved in the column. We could change the separation by changing the polarity of either the column or the mobile phase. Generally it is easier, faster, and cheaper to change the character of the solvent.

The key to changing the separation is to change the *difference in polarity* between the column packing and the mobile phase. Making the solvent polarity more like the column polarity lets compounds elute more rapidly. Increasing the difference in polarities between column and mobile phase makes compounds stick tighter and come off later. The longer a mixture stays on the column, the better the chance for a separation to occur. The effects are more dramatic with compounds having polarities similar to the column.

On a nonpolar column running in acetonitrile, we could switch to a more polar mobile phase, such as MeOH, to make compounds retain longer and have more time to separate. We can achieve much the same effect by adding a known percentage of water, which is very polar, to our starting acetonitrile mobile phase *(step gradient)*. We could also start with a mobile phase containing a large percentage of water to make nonpolar compounds stick tightly to the top of the column and then gradually increase the amount of acetonitrile to wash them off *(solvent gradient)*. By changing either the initial amount of acetonitrile, the final amount of acetonitrile in the mobile phase, or the rate of change of acetonitrile addition, we can modify the separation achieved. Separation of very complex mixtures can be carried out using solvent gradients. There are, however, penalties to be paid in using gradients. More costly equipment is required, solvent changes need to be done slowly to be reproducible, and the column must be reequilibrated to the original starting conditions before making the next injection. *Isocratic* separations made with constant solvent compositions can generally be run in 5–15 min. True analytical gradients require run times of around 1 hr with about a 15-min reequilibration. But some separations can be made only with a gradient. We will discuss gradient development in a later section.

1.1.4 Range of Compounds

Almost any compound that can be retained by a column can be separated by a column. HPLC separations have been achieved based on differences in polarity, size, shape, charge, or specific affinity for a site, and stereo and optical isomerism. Columns exist to separate mixtures of small organic acid present in the Krebs cycle to mixtures of macromolecules such as antibody proteins and DNA restriction fragments. Fatty acids can be separated based on the number of carbons or a combination of carbon number and degree of unsaturation. Electrochemical detectors exist that detect separations at the picogram range for rat brain catecholamines. Liquid crystal compounds have been routinely purified at 50 g per injection. The typical injection, however, is of 20 μl of solvent containing 10–50 ng of sample. Typical runs are made at 1–2 ml/min and take 5–15 min (isocratic) or 1 hr (gradient).

1.2 How Else Could I Get My Separation?

Obviously there are many other analytical tools in the laboratory that could be used to make a specific separation. Other techniques may offer higher sensitivity, less expensive equipment, different modes of separation, or faster and dirtier tools for cleaning a sample before injection into the HPLC. Often a difficult separation can be achieved only by combining these tools in a sequential analysis or purification. I shall try to summarize what I know about these tools, their strengths, and drawbacks.

1.2.1 FPLC—Fast Protein Liquid Chromatography

FPLC, a first cousin of the HPLC, is optimized to run biological macromolecules on pressure fragile agarose or polymeric monobead-based columns. It uses the same basic system components, but with inert fluid surfaces (i.e., Teflon®, titanium, and glass), and is designed to operate at no more than 700 psi. Inert surfaces are necessary since many of the resolving buffers contain high concentrations of halide salts that attack and corrode stainless steel surfaces. Glass columns are available packed with a variety of microporous, high-resolution packings: size, partition, ion exchange, and affinity modes. A two-pump solvent gradient controller, injector valve, filter variable detector, and fraction collector complete the usual system. The primary separation mode is strong anion exchange rather than reverse-phase partition as in HPLC.

FPLC advantages include excellent performance and lifetimes for the monobead columns, inert construction against very high salt concentrations often used in protein chromatography, the capability to run all column types traditionally selected by protein chemists except DEAE, the availability of smart automated injection and solvent selection valves, and very simple system programming. Disadvantages include lack of capability to run high-pressure reverse-phase columns except at very slow speeds, lack of a variable detector specifically designed for the system, and lack of a true autosampler. HPLC components have been adapted to solve the first two problems, but have proved to be poor compromises. Automated valves can partially compensate for the lack of an autosampler.

1.2.2 LC—Traditional Liquid Chromatography

LC is the predecessor of HPLC. It uses a slurry-packed glass column filled with large diameter (35–60 μm) porous solid material. Material to be separated is dissolved in solvent and applied directly to the column head. The mobile phase is placed in a reservoir above the column and gravity fed to the column to elute the sample bands. Occasionally a stirred double chamber reservoir is used to generate linear solvent gradients and a peristaltic pump can be used to feed solvent to the column. Packing materials generally made of silica gel, alumina, and agarose are available to allow separation by partition, adsorption, ion exchange, size, and affinity modes.

A useful LC modification is the quick cleanup columns. The simplest of these is a capillary pipette plugged with glass wool and partially filled with packing material. The dry packed column is wetted with solvent, sample applied, and the barrel is filled with eluting solvent. Sample fractions are collected by hand in test tubes. A further modification of this is the sample filtration and extraction columns (SFE). These consist of large pore packing (30–40 μm diameter) trapped between filters in a tube or a syringe barrel. They are designed to be used with either a syringe to push sample and solvent through the cartridge or a vacuum apparatus to pull solvent and sample through the packed bed into a test tube for collection. Once the sample is on the bed, it can be washed, then eluted in a step-by-step manner with increasingly stronger solvent. These are surprisingly powerful tools for quick evaluation of the effectiveness of a packing material, cleanups, and broad separations of classes of materials. They are available in almost any type of packing available for LC separations: partition, ion exchange, adsorption, and size.

The basic advantages of this technique are low equipment cost and the variety of separation techniques available. Very large and very small columns can be used, they can be run in a cold room, and columns are reusable with careful handling and periodic washing. Disadvantages include relatively low resolving power, overnight runs except for the SFE cartridges, and walking pneumonia from going in and out of cold rooms.

1.2.3 GLC—Gas–Liquid Chromatography

GLC uses a column packed with a solid support coated with a viscous liquid. The volatile sample is injected through a septum into an inert gas stream that carries it into the column. Separation is achieved by differential partition of the sample components between the liquid coating and the continuously replaced gas stream. Eventually each compound flushes off the column and into the detector in reverse order of its affinity for the column. The column is placed in a programmable oven and separation can be modified by using temperature gradients.

Advantages of the technique include moderate equipment prices, capillary columns for high resolutoin, rapid separations, high sensitivity detectors, and direct injection into a mass spectrometer because of the absence of solvents. Disadvantages include the need for volatile samples or derivatives, limited range of column separating modes and eluting variables, the requirement for high purity pressurized carrier gases, and the inability to run macromolecules.

1.2.4 SFC—Supercritical Fluid Chromatography

SFC is a relatively new technique using a silica-packed column in which the mobile phase is a gas, typically carbon dioxide, which has been converted to a "supercritical" fluid under pressure and temperature control. Sample is injected as in a GLC system, carried by the working fluid onto the packed column where separation occurs by either adsorption or partition. The separated components then wash into a high-pressure UV detector flow cell. At the outlet of the detector, pressure is released and the fluid returns to the gaseous state leaving purified sample as a solid. Doping of carrier gas with small amounts of volatile polar solvents such as MeOH can be used to modify the separation.

Advantages of SFC include many of the characteristics of an HPLC separation: high resolving power and fast run times, but with much easier sample recovery. The technique is primarily used as a very gentle method for purifying fragile or heat-labile substances such as flavor oils and perfume fragrances. It is being considered for environmental analysis of water samples. Disadvantages include high equipment cost, the necessity of working with pressurized gases, poor current range of column operating modes and available working fluids, and the difficulty of producing supercritical fluid polarity gradients.

1.2.5 TLC—Thin-Layer Chromatography

TLC separations are carried out on glass, plastic, or aluminum plates coated with thin layers of solid adsorbent held to the plate with an inert binder. Plates are coated with a thick slurry of the solid and binder in a volatile solvent, then allowed to dry before using. Multiple samples and standards are each dissolved in volatile solvent and applied as spots across the solid surface and allowed to evaporate. Separation is achieved by standing the plate in a shallow trough of developing solvent and allowing solvent to be pulled up the plate surface by capillary action. Once solvent has risen a specific distance, the plates are dried and individual compounds are detected by UV visualization or by spraying with a variety of reactive chemicals. Identification is made by calculating relative migration distances and/or by specific reaction with visualizing reagents. TLC can be used in a preparative mode by streaking the sample across the plate at the application height, using nondestructive visualization, scrapping the target band(s) from the plate, and extracting them with solvent. Short (3–4 in.) TLC strips are an excellent quick and dirty tool for checking reaction mixtures and chromatography fractions, and for surveying LC and HPLC solvent/packing material combinations.

Advantages of TLC include very inexpensive equipment and reagents, fairly rapid separations, a wide variety of separating media, visualizing chemicals, and use of solvents and mobile phase modifiers, such as ammonia, not applicable to column separations. Disadvantages include poor resolving power and difficulty in quantitative recovery of separated compounds from the media and binder.

1.2.6 EP—Electrophoresis

EP takes advantage of the migration of charged molecules in buffered solution toward electrodes of the opposite polarity. Electrophoresis separating gels are cast in tube or slab form by either polymerizing polyacrylamide support material or casting agarose of controlled pore size in the presence of a buffer to carry an electrical current. Sample is applied to the gel surface, buffer reservoirs and positive and negative electrodes are connected to opposite ends of the gel, and electrical current is applied across the gel surface. Because electrical resistance in the media generates heat, the gel surface is usually refrigerated to prevent damage to thermally labile compounds. Compounds migrate within the gel in relation to the relative charge on the molecule and, in size-controlled support matrices, according to their size and shape.

Advantages of electrophoresis include relatively low priced equipment, solvents, and media, and very high resolving power for charged molecules, especially biological macromolecules. Disadvantages of EP include working with high-voltage power supplies and electrodes, recovering separated components from a polymeric matrix contaminated with buffer, relatively long separation times in many cases, and the effect of heat on labile compounds.

1.2.7 CZE—Capillary Zone Electrophoresis

CZE is a relatively new technique involving separations in coated capillary columns filled with buffer under the influence of an electrical field. Samples are drawn into and down the column using electrical charge potential. Migration is controlled by the molecule's charge and interaction with the wall coating. Separated components are detected through a fine drawn out transparent area of the column using a variable UV detector. Still under development, CZE offers great potential as improvements are made in injection techniques and in column coatings to add modified partition, size, ion exchange, and affinity capability. Mass spectrometer interfaces are being developed in research laboratories to provide a definitive compound identification.

Advantages of CZE include very high resolving power, fairly short run times, and lack of large quantities of solvent to be disposed. Disadvantages include the fact that this is primarily an analytical tool with little capacity for sample recovery, and that again, there is the necessity of working with relatively high-voltage transformers and electrodes. Resolving variables are limited to column coating, applied voltage, buffer character, strength, and pH.

2

Selecting an HPLC System

Over the years I have encountered a common customer problem when it came to buying HPLC systems. My customers wanted to buy exactly what they needed to get the job done at the very best price. They wanted to be prepared for future needs and problems, but they did not want to buy equipment they did not need or that would not work.

Who should they turn to for advice? The commissioned salesman for the HPLC company obviously could not be considered completely objective. The customer referral from the salesman might be a little better, but who is going to hand out a list of customers who have encountered problems? The local in-house HPLC guru certainly would be more objective, but his information might not be current and his application might be completely different from what they are trying to do. A consultant would be more expensive, probably suffer from the same problems as the guru, and is hard to find without connections to one company or the other.

So, what is the answer? This section I hope! I have no connection to an HPLC company currently, although I have worked for four of the major players in the past. I have taught HPLC extension courses for 9 years and have consulted on a variety of other manufacturer's systems for at least that long. I will try to give you an objective look at the various types of problems that HPLC can solve and my best recommendation for the equipment you will need to solve each one and, at least, a ballpark price (1993 vintage) for each system.

2.1 What Do I Look for in a System?

Like buying computer software, the first step is to decide exactly what you will be using the HPLC for today and possibly in the future. I am not talking about specific separations at this point; those decisions will be used to control column selection, which we will discuss in a moment. What I am really looking for is an overall use philosophy.

2.1.1 Finding a Fit: Detectors and Data Processing

Before we start let me offer some general comments. In the past fixed or filter-variable wavelength UV detectors have been sold with inexpensive systems. Variable detectors were expensive and replacement lamps were expensive and very short lived. This is no longer true, and I would not consider buying an HPLC system without at least a good single-channel variable UV. By the same token, photo diode array UV detectors have been oversold. They have specific applications in method development laboratories, but in their current form they provide useful array information only in a postrun batch mode, not real time. Real time, they are used only as very expensive variable detectors. The computer necessary to extract useful information from the 3D output simply increases their cost.

The other piece of mandatory equipment that has changed recently is the printer–plotter–integrator. Previously, every inexpensive HPLC had to make do with a strip chart recorder. The price differential between the printing integrator and the strip chart has dropped to the point that it does not make sense not to have integrator capability in the laboratory. You may integrate only one out of 10 runs, but when you need it, the capability will be there. The only down side to the integrator is that it commonly uses thermal paper, which does not store well for a permanent record. Often it will be necessary to photocopy the "keeper" chromatograms for further reference.

2.1.2 System Models: Gradients versus Isocratic

There are four basic system types. *Type I* systems are basic isocratic systems used for simple, routine analysis in a QA/QC environment, often for finger-printing mixtures or final product for impurity/yield checking. *Type II* systems are flexible research gradient systems used for both methods development, complex gradients, and dial-a-mix isocratics for routine analysis and standards preparation. They fit the most common need for an HPLC system. *Type III* systems are fully automated, dedicated systems used for cost-per-test, round-the-clock analysis of a variety of gradient and isocratic samples typical of clin-ical and environmental analysis laboratories. *Type IV* systems are fully auto-mated gradients with state-of-the-art detectors used for methods development and research gradients.

2.2 From Whom Do I Buy It?

If you are looking for the name of "the" company from which to buy an HPLC, I am afraid I am going to have to disappoint you. First, that answer is a moving target. Today one company might be the right choice, tomorrow they might have manufacturing and design problems. For one type of system, such as a microbore gradient HPLC, one company may be superior to the competition. For one type of application, such as a biological purification, another company may stand out. Second, HPLC equipment has improved so much that you are fairly safe no matter which hardware manufacturer you select. Service and support have become the differentiating factor, as they really always have been.

2.2.1 Brand Names and Clones

Many of the components that are used in different systems are actually manufactured by a single supplier on an O.E.M. basis. It is still important to buy all your components from the same supplier to prevent a major outbreak of finger pointing in case of problems. If everything comes from one company, they are the ones who are responsible to help solve the problem. Just make sure they are in the business of providing for customer needs. Buying expensive systems does not necessarily guarantee good customer support. Buying from the lowest bidder or buying the cheapest system possible almost ensures that *you* are the customer support. Low margin companies do not have large budgets to plow into support facilities. By the same token, large companies often have so much overhead that little is left for support.

A company's support reputation may change with time and owners. Support is expensive and only the best companies believe in it over the long haul. Find out what a company's reputation for customer support is today from current users.

Your best support will probably come from your local sales and service representatives. If they are good they can help you interface with the company and make sure that problems get solved. Remember that service representatives solve electrical and mechanical, not chemical and column problems. *You* must be able to distinguish between these. With luck, the sales representative will have the proper background and training to be of some assistance in solving these problems. If that training consists of selling used cars, it may not be of much assistance when your column pressure reaches 4000 psi and your peaks have merged into a single mass. Find out how much help your sales representative has been to the person in the lab across the hall.

2.2.2 Hardware—Service—Support

With many laboratory instruments, equipment specifications alone control the decision of which instrument you should buy. However, HPLC systems are so flexible, can run so many types of columns, and have enough control variables, that hardware decisions alone are insufficient in helping you decide which system you need to solve your application problems. I finally designed a diagram to aid in explaining how to buy an HPLC system (see Figure P.1).

If you are buying a water bath for the laboratory you need only consider the temperature range and whether it is UL rated. All you can do is turn it on and set the temperature. Price and hardware considerations are enough to make your decision. If it is critical to your work that the water bath always works, you either buy a backup unit or you buy from a company that will provide excellent and prompt on-site service. At this point, the second leg of the success triangle begins to come into play. In an HPLC system, hardware, service, and support are all critical to guarantee your HPLC success. If you buy from a company that provides only hardware, you must provide service and support. If the company has good hardware and a responsive serviceman, but no support, then you must provide the support. This might mean reading a book and attending courses to become "the HPLC expert." It might mean hiring an HPLC consultant. It might mean getting only a portion of the capability of your system.

The HPLC should be a tool to solve your research problems, not a research problem of its own. Think how much your time is worth. (If you do not know, ask your boss, who knows well!) Selecting a company that can provide excellent hardware, responsive service, and application support after the sale can be one of the most economical decisions you will ever make, no matter what the initial cost of the system turns out to be.

2.3 What Will It Cost?

HPLC companies tend to sell Type II systems when a Type I will do and Type III systems when a Type II would be sufficient for the job. I have tried to estimate a range of prices that I have sold these systems for recently (1993). Inflation will drive these prices up; the very real competition in this field tends to hold prices down. Let us look at each type in turn.

2.3.1 Type I System—QC Isocratic (Cost: $10,000–12,000)

This system is made up of a reservoir, pump, injector, detector, and integrator. The Rheodyne manual injector has pretty much become the standard in the industry. It gives good, reproducible injections, but the fittings used on it are different from any other fitting in the system and are very difficult to connect or disconnect because of tight quarters in the back of the injector. I recommend that you get a variable UV detector as your work horse monitor, and add other monitors as the need arises (i.e., electrochemical detector for catacholamines, a fluorometer for PNAs). The integrator lets you record or integrate. If you dislike working with thermal paper, you can photocopy for long-term storage or look around for a plain paper integrator. Stay with modular systems. Systems in a box are cheaper because of a common power supply, but not nearly as flexible when there are problems with a single component, with upgrading as application, or available equipment change.

2.3.2 Type II System—Research Gradient (Cost: $17,000– 24,000)

The Type II system comes in two flavors. They vary based on the type of gradient pumping system they contain: low-pressure mixing or high-pressure mixing. The rest of the system is the same: injector, variable detector, and printing integrator. Autosamplers allow 24 hr operation, but most university research laboratories find graduate students to be less expensive.

A few years ago I would have always recommended the high-pressure mixing system even though it was more expensive; performance merited the difference. Today, it depends on the applications you anticipate running. If you plan on running 45-min gradients to separate 23 different components, some of them as minor amounts such as with PTH amino acids, then I recommend a dynamically stirred, two-pump, high-pressure mixing system. If, on the other hand, you will mainly be doing scouting gradients, dial-a-mix isocratics, and the occasional uncomplicated gradient, the low-pressure mixing system would be excellent and save you about $4,000. This system has the advantage of giving you three- or four-solvent capability, which would be advantageous in scouting and automated washout, but which *requires* continuous, inert gas solvent degassing. I generally find their gradient reproducibility performance to be about 95% of that of the high-pressure mixing system. Gradient performance from 0 to 5% and 95 to 100% B may be worse than 95% and should be checked carefully before buying (see Chapter 9).

You can replace the integrator with a computer-based data acquisition system if you must, but let the buyer beware. At the moment, most of the computer-based systems are very poor afterthoughts by the HPLC manufacturers. They cost at least $5,000 (1992) not counting the computer, are IBM clone based, and are not very user friendly. The system made by Axxion runs on most systems, is competitively priced, and is reasonably friendly. I am not impressed with most of the control/data acquisition systems I have seen. If I was going to buy any, I would get data acquisition/processing only and wait for performance to improve and price to drop—which they will surely do! My operating rule is to *try before you buy*—and think again. I have been using personal computers for 14 years; I am a fan, but I'm still not convinced. If you do buy, keep your integrator; you will thank me.

2.3.3 Type III System—Automated Clinical (Cost: $25,000– 35,000)

The most common job for these systems is the fast running isocratic separation. They could be built up from the QC isocratic, but dial-a-mix isocratic is faster and more convenient since it switches from job to job. These systems come in the same two flavors as the research gradient: low- and high-pressure mixing.

They also replace the manual injector with an autosampler, allowing 24 hr operation. For thermally labile samples that need to be held for a period of time before being injected, there are autosampler chillers available.

The components in these systems tie together, start with a single start command, and may be capable of checking on other components to make sure of their status. The controllers usually allow different method selection for different injection samples. The more expensive autosamplers allow variable injection volumes for each vial. Since these laboratories must retain chromatograms and reports for regulatory compliance and good laboratory practice, they are moving more toward computer control/data acquisition. At the moment this will add an additional $5,000 to the cost above for software and hardware. This assumes that the computer system replaces the controller and integrator at purchase.

2.3.4 Type IV System—Automated Methods (Cost: $23,000– 46,000)

Another fully automated gradient system, this system is most commonly found in Industrial Methods Development laboratories. They usually have an autosampler, a multisolvent gradient, at least a dual-channel, variable UV detector, and an integrator capable of interfacing with a computer for reports. They may also have a fraction collector to be used in standards preparation. Some laboratories will replace the variable detector with a diode array detector/computer combination that can run the cost of this system to $50,000. Of course, you could have two Type II systems for the same price. It depends on what you are trying to achieve and how heavily budgeted your department is at the moment.

2.4 Columns

The decision about which HPLC column to choose is really controlled by the separation you are trying to make and how much material you are trying to separate and/or recover. I did a rather informal survey of the literature and my customers 10 years ago to see which columns they used. I found 80% of all separations were done on reverse-phase columns (80% were done on C_{18}), 10% were size separation runs, 8% were ion-exchange separations, and 2% were normal-phase separation on silica and other unmodified media, such as alumina. The percentage of size- and ion-exchange separations has increased recently because of the importance of protein purification by size and ion exchange, and the growing use in industry of ion exchange on pressure-resistant polymeric supports.

2.4.1 Sizes: Analytical and Preparative

Columns vary in physical size depending on the job to be accomplished and the packing material used. There are four basic column classes: microbore (1–

2 mm i.d.), analytical (4–4.5 mm i.d.), semipreparative (10–25 mm i.d.), and preparative (1–5 in. i.d.). Column lengths will range from a 3-cm ultrahigh resolution, 3-μm packed Microbore column to a 160-cm semipreparative column packed with 5-μm packing. The typical analytical column is a 4.2-mm i.d. $\times$ 25-cm C_{18} column packed with 5-μm media.

Size separation columns need to be long and thin to provide a sufficiently long separating path. Preparative ion exchange and affinity columns should be short and wide to provide a wide separation surface.

2.4.2 Separating Modes: Selecting Only What You Need

Column decisions should be made in a specific order based on what you are trying to achieve. First, decide whether you are trying to recover purified material or simply analyzing for compounds and amounts of each present (Figure 5.3).

If you are going to make a preparative run, how much material will you inject? Deciding this allows you to select an analytical (microgram amounts) column, a semipreparative (milligrams) column or a preparative (grams) column depending on the amounts to be separated (Table 11.1).

Once the column size issue is decided, the next column decision is based on the type of differences that will separate the molecules. The separating difference might be size, the charge on the molecules, their polarities, or a specific affinity for a functional group on the column.

For size differences, select a size-exclusion or gel-permeation column. A further decision needs to be made based on the solubilities of the compounds. Size separation columns are supposed to make a pure mechanical separation dependent only on the diameter of the molecules in the mixture. Compounds come off in order of size, large molecules first. Solvent serves only to dissolve the molecules so they can be separated. Size columns come packed with either silica-based or gel-based packings in solvents specifically for samples dissolved in either aqueous or organic solvents. Do not switch solvents or solvent types on gel-packed columns; differential swelling can change the separating range, cause column voiding, or even crush the packing.

For charge differences, select either an anion-exchange or cation-exchange column, either gel-based or a bonded-phase silica. Anion-exchange columns retain and can separate anions or negatively charged ions. Cation-exchange columns retain and separate positively charged cations. Silica-based ion-exchange columns are pressure resistant, but are limited to pH 2.5–7.5 and degrade in the presence of high salt concentrations, which limits cleaning contaminants off the column or separation of strongly bound compounds. The functional groups on the column can have permanent charges (strong ion exchangers, either quaternary amine or sulfonic acid) or inducible charges (weak ion exchangers, either carboxylic acid or secondary/tertiary amine). The latter types can be cleaned by column charge neutralization through mobile-phase pH modification. Ion exchangers do not retain or separate neutral compounds or molecules with the same charge as the column packing.

For polarity differences, select a partition column. Look at solubilities in aqueous and organic solvents again. Compounds soluble only in organic solvents should be run on normal-phase (polar) columns. Compounds with structural or stereoisomer differences should be run on normal-phase columns. Most compounds soluble in aqueous solvents should be run on reverse-phase columns. Although C_{18} columns are commonly used, intermediate phase columns, such as the phenyl, C_8, cyano, and diol columns, offer specificity for double bonds and functional groups. Polarity-based separations can be modified by additives to the mobile phase, such as strong solvent changes, pH modification, and ion pair agents.

For specific receptor or shape differences, select an affinity column. The packing surface functional group can be modified to attract and separate only specific compounds or classes of compounds. The attached moiety might be the recognition site for an enzyme, a dye specific for a class of enzymes, an antigen specific for an antibody, or a stereospecific functional group to attract one stereo antimer more tightly than its mirror image. These columns do not retain molecules that fail to fit the specific attachment criteria. These columns are dedicated for this specific use and the column must be modified for its specific function before use.

This selection of separating modes is an oversimplification, but it serves as a good first approximation. There is rarely such a thing as a pure size column or a column that separates solely by partition. Many size columns control pore size by adding bonded phases that can exhibit a partition effect. The underlying silica support can add a cation-exchange effect. A bonded phase column's pore size can introduce a size exclusion effect and its silica surface an ion-exchange effect. Most separations are a combination of partition, size, and ion-exchange effects, generally with one separating mode dominating and others modifying the interactions. This can be a problem when trying to introduce simple changes in a separation, but it can be used to advantage if you are aware that they might be present.

2.4.3 Tips on Column Use

Here are a few *tips on column usage* that will make your life a little easier:

1. Keep the pH of bonded-phase silica columns between 2.0 and 8.0 (better is pH 2.5–7.5). Solvents with a pH below 2.0 remove bonded phases; all silica columns dissolve rapidly above pH 8.0 unless protected with a saturation column.
2. Always wash a column with at least 6 column volumes (approximately 20 ml for a 4 mm $\times$ 25 cm column) of a new solvent or a bridging solvent between two immiscible solvents.
3. Do not switch from organic solvents to buffer solution or vice versa. Always do an intermediate wash with water. Buffer precipitation is a major cause of system pressure problems. You may be able to go from less than 25% buffer to organic and get away with it, but you are forming a very bad habit

that will get you into trouble later on. I usually keep a bottle of my mobile phase minus the buffer on the shelf for column washout at the end of the day. This also can be used for buffer washout, but a water bridge is still the best.

4. Do not shock the column bed by rapid pressure changes, by changes to immiscible solvents, by column reversing, or by dropping or striking the column on the floor or the desk top.

5. Pressure increases are caused by compound accumulation, by column plugging with insoluble materials, or by solvent viscosity changes. It is poor practice to run silica-based columns above 4000 psi (see Chapter 10 on Troubleshooting for cleaning). Keep organic polymer columns and large pore silica size separation columns below 1000 psi or lower if indicated in the instructions supplied with the column. Set your pump overpressure setting to protect your column. Solvents such as water/methanol, water/isopropanol, and DMSO/water undergo large viscosity pressure changes during gradient runs. Adjust your flow rates and overpressure settings to accommodate these increases so the system does not shut down.

6. Use deoxygenated solvents for running or storing amine or weak anionic ion-exchange columns (see Chapter 6, under packing degradation, for a deoxygenation apparatus).

7. Wash out buffers, ion pairing reagents, and any mixture that forms solids on evaporation before shutting down or storing the column. Store capped columns in at least 25% organic solvent (preferably 100% MeOH or acetonitrile) to prevent bacterial growth.

CHAPTER

3

Running Your Chromatograph

This chapter is designed to help you get your HPLC up and running. We will walk through making tubing fittings, putting the hardware together, preparing solvents and sample, initialization of the column, making an injection, and, then, getting information from the chromatogram produced. Let us begin with connecting the hardware and work our way toward acquiring information.

3.1 Setup and Start-up

When your chromatograph arrives someone will have to put it together. If you bought it as a system, a service representative from the company may do this for you. No matter who will put it together, you should immediately unpack it and check for missing components and for shipping damage.

If you bought only components or if you are inheriting a system from someone else, you will have to put it together yourself. More than likely, you will need, at a minimum, a 10-ft coil each of 0.010-in. (ten-thousandths) and 0.020-in. (twenty-thousandths) tubing, compression fittings appropriate to your system, cables to connect detectors to recorder/integrators and pumps to controller, and tools. Our model will be a simple, isocratic system: a single pump, a flush valve, an injector, a C_{18} analytical column, a fixed wavelength UV detector, and a recorder (Fig. 1.4). The first thing we need to do is to get the system plumbed up or connected with small internal diameter tubing. For now, check the columns to make sure they were shipped or were left with the ends capped. We will put them aside until later.

3.1.1 Hardware Plumbing 101: Tubing and Fittings

We will need $\frac{1}{8}$-in. stainless steel HPLC tubing with 0.020-in. i.d. going from the outlet check valve of the pump to the flush valve and on to the injector inlet. Three types of tubing are used in making HPLC fittings: 0.04, 0.02, and 0.01 in. i.d.; the latter two types are easily confused. If you look at the ends of all three types, 0.04 in. looks like a sewer pipe, more hole than tube. Look at the tubing end; if you can see a very small hole and think that it is 0.01 in., it is probably 0.02 in. If you look at the end of the tubing, and, at first glance, think it is a solid rod and then look again and can barely see the hole, it is 0.01-in. tubing. From the injector to the column and from the column on to the detector we will use 4-in. pieces of this 0.010-in. tubing.

It is critically important to understand this last point. There are two tubing volumes that can dramatically affect the appearance of your separation: the ones coming from the injector to the column and from the column to the detector flow cell. It is important to keep this volume as small as possible. The smaller the column diameter and the smaller the packing material diameter, the more effect these tubing volumes will have on the separation's appearance (peak sharpness).

A case in point is a troubleshooting experience that I had. We were visiting a customer who had just replaced a column in the system. The brand new column was giving short, broad, overlapping peaks. It looked much worse than the discarded column, but retention times looked approximately correct. Since the customer was replacing a competitive column with one that we sold, I was very concerned. I asked her if she had connected it to the old tubing coming from the injector and she replied that the old one did not fit. She had used a piece of tubing out of the drawer that already had a fitting on it that would fit. This is *always* dangerous, since fittings need to be prepared where they will be used or they may not fit properly. They can open dead volumes that serve as mixing spaces. I had her remove the column and looked at the tubing. Not only was the end of the tubing protruding beyond the ferrule too short, the tubing was 0.04 in. i.d. This is like trying to do separations in a sewer pipe. We replaced it with 0.01-in. tubing, made new new fittings in the holes they were to connect with, and reconnected the column. The next run gave needle-sharp, baseline-resolved peaks!

To make fittings we need to be able to cut stainless steel tubing. Do not cut tubing with wire cutters; that is an act of vandalism. Tubing is cut like glass. It is scored around its circumference with a file or a microtubing cutter. The best apparatus for this is called a Terry Tool and is available from many chromatography suppliers. If adjusted for the inner diameter of the tubing, it almost always gives cuts without burrs. If you do not have such a tool, score around the diameter with a file. Grasp the tube on both sides of the score with blunt nosed pliers and gently flex the piece to be discarded until the tubing separates. Scoring usually causes the tubing to flare at the cut. A flat file is used to smooth around the circumference. Then, the face of the cut is filed at alternating 90° angles until the hole appears as a dot directly in the center of a perfect circle.

The ferrule should then slide easily onto the tubing. Be sure not to leave filings in the hole; connect the other end to the pumping system and use solvent pressure from the pump to wash them out.

The tubing is connected to the pump's outlet check valve by a compression fitting. The fitting is made up of two parts: a screw with a hex head and a conical shaped ferrule (Fig. 3.1a). The top of the outlet valve housing has been drilled and treaded to accept the fitting.

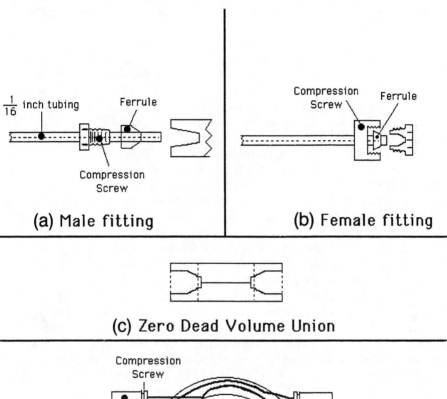

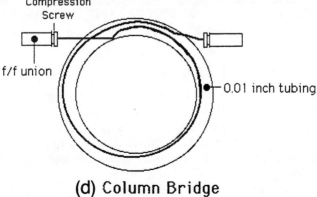

Figure 3.1. Compression fittings. (a) Male fitting; (b) Female fitting; (c) Zero dead volume union; (d) Column bridge.

First the compression screw then the ferrule are pushed on to the tubing; the narrow end of the ferrule and the treads of the screw point toward the tubing's end. The end of the tubing is pushed snugly into the threaded hole on the check valve. Slide the ferrule down the tube into the hole, followed by the compression screw. Using your fingers, tighten the screw until it is as snug as possible; then use a wrench to tighten it another quarter turn. As the screw goes forward, it forces the ferrule against the thread and squeezes it down on to the tubing, forming a permanent male compression fitting. The fitting can be removed from the hole, but the ferrule will stay on the tubing. The tubing must be cut to remove the ferrule.

It is important not to overtighten the fitting. It should be just tight enough to prevent leakage under pressure. Try it out. If it leaks, tighten it enough to stop the leak. By leaving compliance in the fitting, you will considerably increase its working life time. Many people overtighten fittings. If you work at it, it is even possible to shear the head off the fitting. But please, do not.

There is a second basic type of compression fitting, the female fitting (Fig. 3.1b), that you will see on occasion. Some column ends have a protruding, threaded connector tube and will require this type of fitting. This fitting is made from a threaded cap with a hole in the center. It slides over the tubing with its threads pointed toward the tubing end. A ferrule is added exactly as above and the tubing and the ferrule are inserted into the end of the protruding column tube with external threads. Tightening the compression cap again squeezes the ferrule into the tapered end of the tube and down onto the tubing, forming a permanent fitting. The third type of device for use with compression fittings is the zero dead volume union (Fig. 3.1c). A union allows you to connect two male connection fittings.

You will find that stainless steel fittings will cause a number of headaches over your working career. An easier solution in many cases is the polymeric "finger tight" fittings sold by many suppliers such as Upchurch and SSI. These fittings slide over the tubing and are tightened like stainless steel fittings, but are not permanently "swagged" onto the tubing and can be reused. They are designed to give a better zero dead volume fitting, but they have pressure and solvent limits. They are also more expensive, but only in the short run.

3.1.2 Connecting Components

New pumps are generally shipped with isopropanol or a similar solvent in the pump head and this will need to be washed out. Always try and determine the history of a pump before starting it up. Systems that have not been run for a while may have dried out. If buffer was left in the pump, it may have dried and crystallized. In any event, running a dry pump can damage seals, plungers, and check valves.

First we will need to hook up the pump inlet line. This consists of a length of large-diameter Teflon® tubing with a combination sinker/filter pushed into one end and a compression fitting that will screw into the inlet fitting at the bottom of the pump head on the other end. Drop the sinker into the solvent reservoir and screw the other end into the inlet check valve housing.

The next step is to use compression fittings to hook the pump outlet check valve to the flush valve with a length of 0.02-in.-i.d. tubing. A flush valve is a small needle valve used to prime the pump by diverting solvent away from the column when rapidly flushing the pump to atmospheric pressure. Open the valve and the line is vented to the atmosphere. This removes back pressure from the column, a major obstacle when trying to push solvent into a plumbed system.

From the flush valve we can connect with fittings and 0.02-in. tubing onto the injector inlet port. The back of the injector usually has ports for an inlet and an outlet line, two ports for the injection loop, and a couple of wash ports. If a sample loop is not in place, connect it, then make a short piece of 0.01-in.-i.d. tubing with fittings to be used in connecting the column. Use the column end to prepare the compression fitting that will fit into it. At the outlet end of the column, hook up with compression fittings a piece of 0.01-in. tubing that connects to the detector flow cell inlet line. When this is done remove and recap the column and set it aside.

Next we are going to create a very useful tool for working with the HPLC system. I call it a "column bridge" (Fig. 3.1d). It bridges over the place in the system where we would normally connect the column. It is very valuable for running, diagnosing, and cleaning a "columnless system." It is made up of a 5-ft piece of 0.01-in. tubing with a male compression fitting on each end screwed into zero dead volume unions (female/female). Our column bridge now has two ends simulating the end fittings on the column.

Connect one end of our column bridge to the tubing from the injector outlet; the other end is connected to the line leading to the detector flow cell. We have one more line to connect to complete our fluidics. A piece of 0.02-in. tubing can be fitted to the detector flow cell outlet port to carry waste solvent to a container. In some systems, this line will be replaced with small-diameter Teflon® tubing.

In either case, the line should end in a backpressure regulator, an adjustable flow resistance device designed to keep about 40–70 psi backpressure on the flow cell to prevent bubble formation that will interfere with the detector signal. Air present in the solvent is forced into solution during the pressurization in the pump. The column acts as a depressurizer. By the time our flow stream reaches the detector cell, the only pressure in the system is provided by the outlet line. If this is too low, bubbles can form in the flow cell and break loose, resulting in sharp spikes in the baseline. The backpressure regulator prevents this from happening.

The final connections are electrical. A power cable needs to be connected to the pump. Check the manuals to see if fuses need to be installed and do so if required. Finally, connect the 0- to 10-mV analog signal connectors on the back of the detector to similar posts on the strip chart recorder. Connect red to red, black to black. If a third ground wire is present in the cable connect it only at one end, either the detector or the recorder end. (*Note:* The ground wire connects to the cable shield, which is wrapped around the other two wires in the cable. If no ground is connected, no shielding of the signal occurs. If both ends of a ground are connected, the shield becomes an antenna; this is worse than no shield at all.)

Now our system is ready to run. We will need to prepare solvent, flush out each component, then connect, flush out, and equilibrate the column before we are ready to make our first injection of standard.

3.1.3 Solvent Cleanup

Before we tackle the column, let us look at how to prepare solvents for our system. I have found that 90% of all system problems turn out to be column problems. Many of these can be traced to the solvents used, especially water.

Organic solvents for HPLC are generally very good. There are four rules to remember: always use HPLC grade solvents, buy from a reliable supplier, filter your solvents, and check them periodically with your HPLC. Most manufacturers do both GLC and HPLC quality control on their solvents; some do a better job than others. The best way to find good solvents is to talk to other chromatographers.

Even the best solvents need to be filtered. I have received solvents, from what I considered to be the best manufacturer of that time, that left black residue on a 0.54-μm filter. There is a second reason to filter solvents. Vacuum filtration through a 0.54-μm filter on a sintered glass support is an excellent way to do a rough degassing of your solvents. Because of their filter and check valve arrangements, some pumps cavitate and have problems running solvents containing dissolved gases.

There are numerous filter types available for solvent filtration. Cellulose acetate filters should be used with aqueous samples with less than 10% organic solvents. With much more organic in the solvent, the filter will begin to dissolve and contaminate your sample. Teflon® filters are used for organic solvent with less than 75% water. The two types are easily told apart; the Teflon® tends to wrinkle very easily, while the cellulose is more rigid. If you are using the Teflon® with high percentages of water in the solvent, wet the filter first with the pure organic solvent, then with the aqueous solvent before beginning filtration. If you fail to do this it will take hours to filter a liter of 25% acetonitrile in water. Recently, nylon filters for solvent filtration have appeared that can be used with either aqueous or organic solvents. They work very well as a universal filter, but use with very acidic or basic solutions should be avoided.

If you are still having problems after vacuum filtration, try placing the filtrate in an ultrasonication bath for 15 min (organic solvents) or 35 min (aqueous solvents). Ultrasonic baths large enough to accept a 1-liter flask are in common use in biochemistry laboratories and are very suitable for HPLC solvent degassing. Stay away from the insertion probe type of sonicator; they throw solvent and simply make a mess. Ultrasonication is much better than heating for removing gases from mixed solvents. There is much less chance of fractional distillation with solvent compositional change when placing mixtures in an ultrasonic bath. One manufacturer actually designed an HPLC system that was designed to remove dissolved gas by heating under a partial vacuum. Obviously they never used rotary vacuum flash evaporators in their laboratories, at least not intentionally!

Other techniques recommended for solvent degassing involve bubbling gases (nitrogen or helium) through the solvent. Helium sparging is partially effective, but expensive when used continuously. It is required in some low-pressure mixing gradient systems as will be described later. The only other time I use any of these degassing techniques is in deoxygenating solvent for use with amine or anionic-exchange columns, which tend to oxidize (see Fig. 6.3).

Water is the major offender for column contamination problems. I have diagnosed many problems, which customers initially blamed on detectors or pumps or injectors, that turned out to be due to water impurities. Complex gradient separations are especially susceptible to water contamination effects.

In one case, a customer was running a PTH amino acid separation, a complex gradient run on a reverse-phase column. He would wash his column with acetonitrile, then water, and run standards. Everything looked fine. Five or six injections later his unknown results began to look weird. He ran his standards again only to find the last two compounds were gone. He blamed the problem on the detector. I said it looked like bad water. He exploded, and told me that his water was triple distilled and good enough for enzyme reactions. It was good enough for HPLC, he said. Over the following 6 months we replaced every component in that system. Eventually, the customer borrowed HPLC grade water from another institution, and washed his column with acetonitrile, then with water. The problem disappeared and never came back—until he went back to his own water. Nonpolar impurities codistilling with the water were accumulating at the head of the column and retaining the late runners in the column.

While HPLC grade water is commercially available, I have found it to be expensive and to have limited shelf life. The best technique for purifying water seems to be to pass it though a bed of either reverse-phase packing material or of activated charcoal, as in a Milli-Q system. Even triple distillation tends to codistil volatile impurities unless done using a fractionation apparatus.

I have used an HPLC and an analytical C_{18} column at 1.0 ml/min overnight to purify a liter of distilled water for the next day's demonstration run. The next morning, I simply washed the column with acetonitrile, then with water, equilibrated with mobile phase, and ran my separation. It might be a good idea to reserve a column strictly for water purification.

An even better solution is to use vacuum filtration through a bed of reverse-phase packing. Numerous small C_{18} SFE cartridges are available that are used for sample cleanup and for trace enrichment. They are a tremendous boon to the chromatographer for sample preparation, but also can be of help in water cleanup. These SFE cartridges are a dry pack of C_{18} packing and must be wetted before use with organic solvent, then with water, or an aqueous solution. You wash first with 2 ml of methanol or acetonitrile and then with 2 ml of water before applying sample. If you forget and try to pass water or an aqueous solution through them, you will get high resistance and nonpolars will not stick. SFE cartridges contain from 0.5 to 1.0 g of packing and will hold approximately 25–50 mg of nonpolar impurities. If care is taken not to break their bed, they can be washed with acetonitrile and water for reuse. Eventually, long eluting impurities will build up and the SFE must be discarded. I have used them about six times, cleaning about a liter of single distilled water on each pass. If larger quantities of water are required, there are commercially available vacuum cartridge systems using large-pore, reverse-phase packing designed to purify gallons of water at a time.

The most common choice for large laboratories is mixed-bed, activated charcoal and ion-exchange systems that produce water on demand. These systems usually have a couple of ion-exchange cartridges and one activated charcoal filter in series. They work very well, but I prefer to have the charcoal as the last filter in the purification bank. After all, we are trying to remove organics. I find that the ion-exchange resins break down after about 6 months and begin to appear in the water. The system uses an ion conductivity sensor as an indicator of water purity, but water that passes this test often is still unsuitable for HPLC use.

3.1.4 Water Purity Test

The final step is to check the purity of the solvents. Again I have found the C_{18} column to be an excellent tool for this purpose. Select either 254 nm or the UV wavelength you will be using for the chromatogram. Wash the column with acetonitrile until a flat UV baseline is established and then pump water though the column at 1.0 ml/min for 30 min. This allows nonpolar impurities to accumulate on the column. The final step is to switch back to acetonitrile. I prefer to do this by running a gradient to 100% acetonitrile over 20 min. If no peaks appear after 5 min at final conditions, the water is good. The chromatogram (Fig. 3.2) gives you an idea of the expected baseline appearance.

Peaks that appear during the first acetonitrile washout are ignored as impurities already on the column. Watch the baseline on switching to water. At 254 nm, the baseline should gradually elevate. If instead it drops, you may have impurities in your acetonitrile. If the baseline makes a very sharp step up before leveling off, you may have a large amount of polar impurities in the water. Polar impurities probably will not bother you on reverse-phase columns, but might have some long-term accumulation effects. Peaks appearing during the acetonitrile gradient come from nonpolar impurities in the water that accumulated on the column and are now eluting.

I have done this with water from a Milli-Q system in need of regeneration. Even though their indicator glow light shows no evidence of charged material being released from the ion exchanger, peaks that will affect reverse-phase chro-

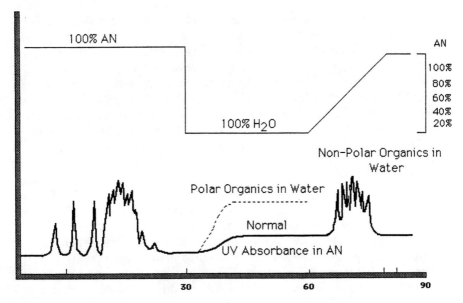

Figure 3.2. Water purity test.

matography show up at around the 70% acetonitrile portion of the gradient run.

If your water passes this test at the wavelength you will be using for your chromatography, you are ready to use it to equilibrate the column. The next step is to flush out the dry system and prepare to add the column.

3.1.5 Start-up System Flushing

Fill the solvent reservoir with degassed, filtered solvent by pouring it down the wall of the flask to avoid remixing air into it. I usually start pumps up with 40–50% methanol in water. Even if the pump was shut down and allowed to stand and dry out in buffer, there is a good chance this will clear it. It is also a good idea to loosen the compression fitting holding the tubing in the outlet check valve at the top of the pump head to relieve any system backpressure. This is an especially important step to use if the column is still connected. When running with a column bridge, as we are, it is less important.

The first step is to ensure that the pump is primed. This may mean pushing solvent from an inlet manifold valve through the inlet valve and into the pumping chamber. A few pumps on the market, like the old Waters M6000, use spring-loaded check valves, so you may have to really work to get solvent into the chamber. With other pumps, you open a flush valve and use a large priming syringe to pull solvent through the pumphead. The next step is either to turn the pump flow to maximum speed or use the priming function of the pump, which does the same thing.

As soon as the pump begins to pump solvent by itself, tighten the outlet compression fitting and drop the flow rate to about 1 ml/min. The pump is ready to run and should be allowed to pump into a breaker for a few minutes to wash out any machining oils, if new, or soluble residues or dissolved buffers if old.

Before we move on, let us talk about shutting down a pump. The pump seal around the plunger is lubricated by the contents of the pumping chamber. There is always a microevaporation through this seal/plunger combination, whether the pump is running or not. Buffers and other mobile phases containing dissolved solids should not be left in a pump when it is to be turned off overnight. This evaporation causes crystallization on the sapphire plunger and can result in either plunger breakage or seal damage on starting up the pump. Solvents containing dissolved solids should always be washed out before the pump is shut down. I prefer to wash out and leave a pump in 25–50% methanol/water to prevent bacteria growth in the fluidics system.

Occasionally, I have had to leave buffer in a pump overnight. When I do that I leave the pump running slowly (0.1 ml/min) and leave enough solvent in the reservoir so that it can run all night. This has an additional value of washing the column overnight. If the column is clean and does not require further washing, you can throw the detector outlet into your inlet reservoir and recycle the solvent, ensuring that you will not run out.

Now we can move past the flush valve to the next major system component, the injector. Whichever position you find the injector handle in, leave it there! *Never turn the handle on a dry injector.* The injector seal is hardened Teflon® facing a metal surface and can tear if not lubricated with solvent. Once solvent is flowing through the injector to lubricate the seal, turn the handle to the inject position so that the sample loop is washed. Watch the pressure gauge on the pump; a plugged sample loop will cause the pressure to jump. If this happens go to the troubleshooting section in Appendix C.

3.1.6 Column Preparation and Equilibration

The next step is to hook up the column. Stop the pump flow. I assume you have a C_{18} column compatible with 40% methanol/water (otherwise, select a solvent appropriate for your column). Disconnect the column bridge, remove the column fittings from both ends of the stored column, and connect the inlet end to the line coming from the injector. The inlet end of a column is almost always marked; check for an arrow or a tag pointing in the direction of flow. I have always preferred to hook up a column with some solvent running. Turn the flow rate on the pump to 0.2 ml/min. Fill the end of the column with solvent and screw in the compression fitting at the end of the injector line. Place a beaker at the outlet end of the column to catch washout solvent. Flush the column with start-up solvent if it is an old column that might have been stored in buffer. (This is a very bad technique, but you never know if you were not the last person to use the column! It is a good idea to label a column with the last solvent used before you store it.)

Next, change the solvent in the reservoir to 70% acetonitrile in water, turn the pump on, and flush it with the new solvent. Turn the flow rate up to 1.0 ml/min while catching the column effluent in a beaker. Check back up line for leaks; if you see any, tighten the appropriate fittings until the leaks just stop. You will always have leaks! If you do not you are probably overtightening your fittings. Leaks are messy, but are probably a sign of successful technique (leaks, not streams).

Check the pump pressure. The pump pressure gauge and the baseline trace are the two major tools for diagnosing system problems. If the column was shipped in isopropanol or methanol it should start high (3000–4000 psi) then slowly drop to around 2000–3000 psi.

Stop the pump flow and connect the column outlet with the short piece of 0.10-in. tubing connected to the inlet of the detector flow cell. Resume flow to the column. Turn the detector on and start the recorder chart speed at 0.5 cm/min. You should have a flat baseline. If the baseline continues to drift up or down, the column still has not finished its washout and equilibration, or the detector is not warmed up.

By the way, I must hasten to add that we really have not reached a true equilibration at this point. I have been informed by the experts that it takes about 24 hr to reach a true equilibration on reverse-phase packings. However, after six column volumes we have generally reached a reproducible equilibration point good enough for our purposes.

We are now ready to prepare for injecting a sample. Let us turn our flow rate down to 0.1 ml/min and get our sample ready.

3.2 Sample Preparation and Column Calibration

The worst thing a chromatographer can do is to grab a column out of its box, slap it into an HPLC, and shoot a sample. Before we begin, it is important to make sure the sample is clean. We will talk about removing soluble contaminants later. Here we are going to be dealing with suspended solids or particulates. Second, we need to know the initial condition of the column, so that we may return to it when we begin to develop problems. In other words, we need to do column quality assurance, or QA.

3.2.1 Sample Cleanup

The generally recommended procedure for cleaning samples is to filter them through a 0.54-μm filter in a Sweeny filter holder or a disposable plastic filter cartridge. The same types of filter materials are available as those that were discussed in the section on solvent filtration: Teflon®, nylon, and cellulose. In-line filters are now available that fasten between the syringe barrel and the injection needle. These are useful if you are not sample limited or are doing repeat injections of the same material. I have found that most chromatographers will not

bother with the time, cost, and sample loss that this entails, although I am finding an increase in the use of syringe in-line filters.

Sample clarification is, however, important! The column frit pore size is
usually 2.0 μm; anything larger builds up and plugs the frit. Being a lazy chromatographer, but not a stupid one, I decided to use a different clarification procedure. I place the sample in a microcentrifuge tube and sediment solids by
spinning at maximum speed in a clinical centrifuge (1000g) for 1–2 min. I pull
a sample carefully from the supernatant and shoot that as my sample. It has
the advantage of spinning down most of the solids, can be used on a number
of samples at the same time, works even with very small samples, and is fast
and inexpensive, if you already have the centrifuge. While it may not be as efficient as filtration, most chromatographers are willing to use it on every sample.
It greatly extends column life between cleanups.

A third alternative combines the two techniques. A commercially available
filter/reservoir fits in a microcentrifuge tube. The sample in the reservoir is filtered by spinning the unit. It is more efficient than simple centrifugalization,
but takes longer to assemble and costs more.

Like the oil filter advertisement says, "you can pay me now, or pay me
later." If you don't take time to remove particulates, you will spend much more
time and effort cleaning the column. The choice is yours.

3.2.2 Plate Counts

Once the shipping solvent is washed out of the column, it is important to determine whether the column bed survived shipping and to determine its running
conditions. Most good chromatography laboratories have established a quality
control test for entering columns. A stable test mixture of known running characteristics has been prepared and stored to test new columns.

One commercially available standard used for testing C_{18} columns is a solution of acetophenone, nitrobenzene, benzene, and toluene in methanol. (Many
chromatographers like to add a basic component, such as aniline, as a check
against tailing problems, but these standards degrade on standing in oxygen-
containing solvents.) To adjust for extinction coefficient differences, add 10 μg
of each of the first two ingredients and 30 μg of the last two compounds in 2 ml
of MeOH. Inject 20 μl of the mixture into the column equilibrated with 70%
MeOH in water and read at 254 nm on the UV detector. This is a convenient
mixture since separations between pairs of peaks double as you go to larger
retention volumes. Be sure to keep this mixture tightly stoppered. The last two
compounds will selectively evaporate from the mixture on access to air. For
use at low UV wavelengths, dissolve these same four ingredients in acetonitrile
and run in 60% acetonitrile in water.

Using this or similar mixtures, inject a sample into an equilibrated column,
elute the resolved bands, and record them on the strip chart recorder. Calculate
plate counts for the first and last peak using the "5σ" method discussed in section 4.1.1. Log these numbers in the form V4/V1 = 1.1/6.5; N4/N1 = 7500/

3600. When we see changes in a separation we have been running, we can ree-quilibrate in 70% MeOH/water and rerun our standards. Changes in these ratios will be useful in troubleshooting column problems later on.

Obviously, this mixture will not be useful on other types of columns, although I have used this mixture on C_8 columns. Each column type should have its own standards. They should be stable against both chemical and bacterial changes. With them you always have a touchstone to return to in case of problems.

3.3 Your First Chromatogram

Now that we have our system set up and the column equilibrated and standardized, we are ready to carry out an HPLC separation on a real sample. We might add an internal standard (if necessary, to correct for injection variations), then dilute our sample to a usable concentration and prepare it for injection. After injection, we will record the chromatogram, making sure that it stays on scale. Then, from the trace we obtain, we will calculate elution volumes either by measuring peak heights or by calculating peak areas by triangulation.

We can compare these values of areas or peak heights with known values for standard compounds. From elution volumes or retention times, we can begin to identify compounds. Comparing peak areas or heights to those derived from standard concentrations, we can calculate amounts of material under each peak.

3.3.1 Reproducible Injection Techniques

From the last section, it becomes obvious that we must first make a decision about what we are trying to accomplish. Are we doing scouting, trying to identify compounds by their retention times? Or are we trying to quantitate peaks by comparison to standards?

In scouting, we may be running very expensive samples and have simply to guess at the amount to inject. In this case, I would pull up $> 10 \ \mu l$ of the supernatant in a 25-μl syringe, turn the syringe point up, and pull the barrel back far enough so I could see the meniscus just below the needle. I would check for bubbles at the face end of the barrel, on the inside wall, and at the meniscus. Small bubbles generally can be dislodged by gently snapping the outside wall with your finger. Slowly push the barrel forward to the 10 μl mark, then quickly wipe the outside of the barrel past the tip with a tissue. Place the syringe into the injector syringe port, make sure the injector handle is in the LOAD position, and slowly push the sample into the loop to ensure that the sample goes in as a plug.

If the syringe is new or dry, you may find a large, tenacious bubble clinging to the barrel face. This can often be avoided by rinsing the barrel with the stronger solvent and then with the sample solvent. It can be dislodged by rapidly expelling the sample from the syringe back into the tube (try not to resuspend the pellet) and then slowly pulling up a new sample. Repeat the check for bubbles, expel the excess sample, and wipe before injecting. Do not let the tissue linger at the tip; it can wick up solution from the needle and give irreproducible sampling.

When working with sample we do not mind wasting, the simplest way to achieve reproducible injections is to overfill the loop. With a 20-μl loop we need to flush with at least 30 μl of sample solution to ensure complete removal of mobile phase. Almost all autosamplers operate on the principle of overfilling the sample loop to achieve a reproducible injection.

It is possible to inject partially a loop below its capacity. It is important not to add more than 75% of loop capacity and to inject the sample slowly. Rapid injection can lead to a phenomenon known as "viscous fingering" in which the sample does not cleanly displace solvent out the other end of the loop as a plug, but some of the sample overruns the loop and is lost.

Quantitative sample injection is handled a little differently. We usually know the expected concentration level and retention times. After clarification, we add a known amount of the sample solution and an internal standard to a volumetric flask and dilute. The sample is pulled into the syringe for injection as above.

Internal standards are used for many reasons in chemistry. Here we are using it to correct for differences in sampling volumes. It takes much practice for a person to deliver accurately the same size sample every time if they are partially injecting a loop. It is nearly impossible for two people to deliver accurately the same sample each time. If we add a known amount of internal standard to both our sample and our known standard mixture, we can calculate peak heights or areas relative to that of the internal standard. Variations in the injection size of the sample do not affect these relative areas.

To make the injection, we turn the handle of the injector to the LOAD position (see Fig. 9.9). Push the syringe needle into the needle port and slowly push the barrel forward so the sample goes in as a plug. Leave the needle in the injection port to prevent syphoning of the sample out the waste port. The handle is thrown quickly to the INJECT position. This last step is done quickly to prevent pressure buildup while the ports are blocked in shifting from one position to the other. Remember: load slowly, inject quickly.

Mark the injection point on the chromatogram. Some detectors or integrators will do this automatically if they are connected to a contact sensor on the injector or autosampler. It is good laboratory practice to annotate the chromatogram at the first injection point of the day with the operator's initials, time, date, sample ID number and injection volume, mobile phase composition, flow rate, detector wavelength and attenuation, and chart speed. If a gradient is being run, mark the starting composition, gradient start and end, and final composition. You can annotate later injections only with condition changes, such as sample i.d. and injection size. If you tend to cut your chromatograms apart, however, you may lose critical information if you do not annotate every run with full information. There are commercially available rubber ink pad stamps that provide spaces for all the necessary information. Do not rely on your memory to come up with the data at some future time.

3.3.2 Simple Scouting for a Mobile Phase

The first step is to determine a starting point. If I am handed a mixture of a completely unknown nature, I will probably first try to get more information. I will try to determine the mixture's solubility in organic solvents, the effect of acid on the solubility, and something about the molecular weights and isoelectric points, if it is a mixture of proteins.

If this information is not available, I will try to separate the mixture using a C_{18} column in acetonitrile and water. Something like 70% of the separations in the literature are now made on a C_{18} column. Acetonitrile is my solvent of choice because of its low wavelength transparency, its polarity, and its intermediate position between methanol and tetrahydrofuran. Generally, I will use 254 nm for the detector because the majority of the literature separations can be made at that wavelength.

If I know that the compound is not soluble in aqueous solvents, I will probably select a silica column and a chloroform/hexane mobile phase. Separations of proteins will take me first to a TSK-3000sw column and a 100 mM Tris-phosphate pH 7.2 mobile phase unless I am separating soluble enzymes; then I use a TSK-2000sw column.

For illustration purposes, we will take the most common case. We will start with a 15-cm-long C_{18} column, 254 nm, and acetonitrile/water in a scouting gradient. Scouting gradients are run much more rapidly than analytical gradients. A mixture of the compounds to be separated is dissolved in 25% acetonitrile in water. A sample is injected into an HPLC equilibrated in the same mobile phase and a 20 min gradient is run to 100% acetonitrile.

Examination of the chromatogram while the separation is occurring lets us select conditions for a starting isocratic run. Since we were running very rapidly, conditions inside the column were not at equilibration. We use the gradient position of the first peak maximum as a guide to an isocratic mobile phase. Find the solvent composition from the controller %B output corresponding to the first peak and drop back to 10% less acetonitrile for a 25-cm column (7% less for our 15-cm column). Using the gradient controller to dial-a-mix the solvent, we reequilibrate the column for 15 min at this acetonitrile concentration and reinject our standards.

If all the peaks are accounted for and separated, we have our conditions. If not, we can do k' development, control pH by buffering, or change the stronger solvent or the type of column to produce an α change. We have a starting point, and that is half the battle.

If you do not have a gradient, I have developed a fast isocratic scouting technique. You select the same column and detector wavelength, but equilibrate in the column in 80% acetonitrile in water for our first injection. Strong solvent is selected to blow everything off quickly. Look at the peaks; if they are resolved, quit. If they are still unresolved, mix the mobile phase with an equal volume of water making 40%, reequilibrate, and shoot again. This time the peaks should be much farther apart. If not, do another equal volume dilution to 20%, reequilibrate the column, and reinject the sample.

If the first peak from the 40% run takes more than 20 min or the peaks are too far apart, wash everything off with 100% acetonitrile. Mix mobile phase 80 and 40% in equal volumes to get 60%, reequilibrate, and shoot again. I usually find that I have my conditions by the third run or I need to make a solvent alpha change.

Silica columns are run the same way. Start gradients at 25% chloroform/hexane and run to 100% in 20 min. For isocratic scouting, start at 80% chloroform/hexane and make dilutions with hexane.

3.3.3 The Chromatogram: What Do You Look for?

I usually run scouting samples at an initial UV attenuation of 0.2 AUFS (absorbance units full scale) or refractive index attenuation of $8\times$. This way I can increase attenuation if the peaks start to go off scale or decrease attenuation if they are too small. An integrator will see everything from the baseline up to full attenuation, but you have to be reasonably close if you are using a strip chart recorder or you will lose peak information.

I would rather blow my first sample off scale and have to dilute the second one. At least I know I got the sample in and what the next step should be. If I shoot too little, I wait and wait for something to happen and waste a lot of valuable time. Besides, I have found that the first shot of the day is usually a "column tranquilizer." It seldom agrees with other sample of the day. The second and third agree, but not necessarily with number one. I have discussed this problem with other chromatographers and many have observed the same thing. If this bothers you, remember that chromatography is still art as well as science. Shoot the first sample and go and have some coffee. Then, you can get down to work.

I am often asked if peak heights or peak areas give more accurate results. The answer to this question is yes. When working with mixtures of pure compounds with very little overlap, peak areas give more accurate results. However, my clinical friends, who must quantitate on peaks from complex mixtures with overlapping peaks, insist that peak heights are more accurate.

3.3.4 Basic Calculation of Results

In peak height measurements, we measure the vertical displacement from the baseline and compare that to the peak height of a known standard amount. Peak area calculations are a little more complicated. They are usually done by triangulation; assume a right triangle and multiply the peak height times the half peak width. The areas of each peak are summed to give a total area. Dividing this into the area of each peak gives a relative area percentage for each peak. Like peak heights, peak areas can be compared to peak areas for known compounds to allow calculation of the amounts of compound present.

Another more accurate method is to copy the chromatogram, cut out the peaks, and weigh them. Of course, if you have an integrator, it will do the job for you. They usually can be set to do either peak heights or areas. They also can be calibrated for standard runs and will calculate actual amounts relative to these earlier runs. Some also can be calibrated with compound names to provide annotated output.

Integrators are designed to make the chromatographer's life easier, but they can complicate it if not properly used. They usually have an autozero function that, when selected, looks at the baseline before injection and sets various baseline parameters. This is designed to prevent integration of very small or extraneous peaks, or of baseline noise. On most integrators, autozero must be requested by the operator and should be used every time a detector attenuation change is made. Be aware that you are letting a machine make decisions for you. It is possible to override the machines, and, sometimes, it is possible for you to produce a more accurate repetitive analysis by doing so. We will touch on integrator optimization in Chapter 14 on data collection.

Once we have returned to the baseline from one chromatogram, we are ready to make our next injection. When we have finished for the day, shut off the detector (lamps have finite lifetimes) and the strip chart recorder paper drive. If we are pumping solvents containing solids, they must be washed out before shutting down the pump. The system can be stored overnight or over a weekend with solvent containing more than 50% organic in the mobile phase. If you will be storing longer than a weekend, wash the system out with acetonitrile, remove and cap the column, and store it in its box labeled with the solvent and the last sample run in it.

Part II
HPLC Optimization

4

Separation Models

Three main modes of separation are used in HPLC systems. Partition separation makes up the majority, followed by size separation, and, finally, by ion exchange.

4.1 Partition

Separation in the column occurs when the sample in the mobile phase begins to interact with the stationary packing material. The actual mechanisms for these interactions are still being investigated. They probably involve forces generated by ordering of the charge density separations in polarized compounds such as water. Because of the electron attraction of the oxygen and the bonding angle of the hydrogens with the oxygen, the water molecule exhibits a negative and a positive end. Water achieves a minimum energy state when it can align positive ends toward negative ends. This ordering leads to an "attraction" of polarized molecules for each other.

When nonpolar molecules or portions of molecules are introduced into a polar matrix, the nonpolars will orient in a manner leading to maximized polar interactions. This results in nonpolars being pushed together. The net effect is that *likes attract likes*—polars with polars and nonpolar molecules with nonpolars. Compounds with similar polarities are attracted to each other. The packing material has differential attractions for different compounds in the sample depending on their degree of polarization. Since the mobile phase is continuously being replaced, each component is washed off at a different rate.

Even small differences in attraction for the packing surface, when repeated many thousands of times, lead to a separation.

A good model for this partition is the separation that takes place in a separatory funnel, as we discussed in Chapter 1. If you dissolve a mixture of two components (A and B) in a separatory funnel containing two immiscible liquids, an equilibrium is established for both compounds in each solvent.

If sufficient polarity differences exist between the compounds, each compound will tend to concentrate in the solvent with a similar polarity. Like attracts like. The more polar compound concentrates in the polar layer. The other member of the mixture is forced toward the nonpolar layer. The bottom layer can then be drawn off, taking with it one of the two components. The second component remains behind separated in the upper layer, which could be recovered next.

In our example, we separated a purple mixture made up of a polar red dye and a nonpolar blue dye. Adding this mixture to a separatory funnel containing water and hexane and shaking vigorously will produce two colored layers. The upper (hexane) layer contains the blue (nonpolar) dye; the lower (water) layer attracts the red (polar) dye. The water layer containing the red dye can be drawn off by opening the stopcock (Fig. 1.2). Evaporation of the water will yield the more polar red dye. In a similar manner, we can recover the blue dye from the hexane layer left in the separatory funnel.

The same type of separation occurs in the HPLC column. Either the mobile or stationary phase is polar and attracts the more polar component in the injected mixture. Let us assume that our column packing is polar and we are pumping a nonpolar mobile phase down the column. Both components have a partition affinity for the packing and will be retained. But the more polar of the two will be retained longer. Since equilibration is continuously being upset in favor of the moving liquid phase, the less polar component washes out as the faster moving band and is eluted first from the column (Fig. 1.3). Eventually both compounds will wash off in turn from the column into the detector.

A number of HPLC partition columns of differing polarities are available and will be discussed later. For now, let us consider the example of a separation on a polar, hydrated silica gel column using methylene chloride in hexane as our nonpolar mobile phase.

Silica gel is hydrated silicic acid with a controlled amount of water of hydration. Each silica on the surface of the packing has one or more hydroxyl group associated with the water of hydration. The available proton on the hydroxyl group gives silica its acid nature and along with the hydration shell makes it a very polar surface.

The same purple mixture that we separated in our separatory funnel example is dissolved in the methylene chloride and shot onto the column through the injector. The two compounds to be separated are swept together onto the column. As fresh mobile phase causes them to pass down the column, the more polar component (the red dye) is more highly attracted to the polar column surface and is retained more than the more nonpolar blue dye. The blue dye moves a little faster and begins to pull apart from the red. Finally the blue dye reaches the end of the column and begins to elute into the detector. The detector signals the concentration changes to the strip chart recorder as a voltage change. As the band center of the peak (B) passes the detector, the strip chart recording goes through a maxima, then returns to the baseline (Fig. 4.1). Next, the red dye completes its trip down the column and begins to elute. It also enters the detector and produces a broader peak (A) on the recorder. The peak broadening occurs because the red dye spent more time in the column, giving diffusion more of a chance to spread it out.

Let us now examine the chromatogram produced by this separation. Starting at the point of injection, we follow the baseline to the first deflection. After about 2 min, we see a small positive deflection immediately followed by a small negative deflection, which then returns to the baseline. The center of this peak complex is called the *void volume* (V_0). It represents the amount of mobile phase contained inside the column, but outside the packing material. It is the mobile phase volume necessary to wash out the sample solvent.

This peak occurs because the solvent composition used to dissolve the sample often differs somewhat from the mobile phase. When this sample solvent volume reaches the detector (at V_0), a refractive index difference upset of the baseline occurs. This V_0 peak is very important because it allows us to correct the separation parameters for variations in column lengths. It also gives us some assurance that the sample was actually loaded into the injector and onto the column.

The next peak is that produced by the blue dye (B); we will measure the mobile phase volume at the center of the peak and call it V_B. In the same manner, we can calculate V_A as the retention volume for the red dye. We could measure the distances V_0, V_A, V_B just as easily in minutes since injection or as cen-

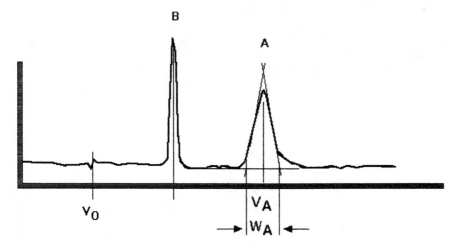

Figure 4.1. Separation model chromatogram.

timeters of graph paper. As we will see, the separation parameters are dimensionless. Thinking in mobile phase volumes eliminates the necessity of considering strip chart and pumping speeds. In the literature you may see V_B described as tr,B the retention time of B, or as the retention length of B in centimeters. These are both referred to as V_B.

4.1.1 Separation Parameters

From these volumes, we can calculate three factors, k', α, and N (see Fig. 4.2), which will then be used to describe a resolution equation (Fig. 4.3). This equation predicts the effect of variations in these factors in controlling resolution within the HPLC column. They are presented here to discuss the variables controlling each of them, their limits, and how you can use them to achieve your separations in a rational manner.

They also serve as a common language when discussing separation problems. With these quantities in hand, it is generally unnecessary to detail other operating conditions. Finally, their most important use is as a diagnostic tool for column problems.

The first factor, the retention factor (k'), is a measure of the relative retention of each peak on the column. In our example, k'_B, the retention factor for the blue peak, is determined by dividing the difference between V_B and V_0 by the void volume. It effectively tells us how long it takes the center of peak B to come off the column relative to V_0. We can derive a similar factor for the red dye (k'_A) or for any peak in a multipeak mixture.

The next factor, the separation factor (α), represents the relative separation between two peak centers on a chromatogram. It is defined as the retention factor of the longer retaining peak divided by the retention factor of the faster peak. Any pair of peaks in the chromatogram will have their own α.

The final factor, the efficiency factor (N), measures the degree of sharpness of a given peak. It is determined by the retention volume of the peak (i.e., V_B) divided by the peak width. Two widths are commonly used for this calculation, the width at one-half the peak height and the width 10% up the peak, the 5σ width. The half-height width is easier to measure, but corrects poorly for peak tailing. Efficiencies calculated from it are optimistically high and are unresponsive to column changes.

$$K' = (V_A - V_O)/V_O$$

$$\alpha = (V_A - V_O) / (V_B - V_O)$$

$$N = 16(V_A/W_A)^2 = 5.42 \, (V_A/W_{.5})^2$$

$$HETP = L/A$$

Figure 4.2.   Separation factors.

$$R = \frac{1}{4}\left(\frac{\alpha - 1}{\alpha}\right)\left(\sqrt{N}\right)\left(\frac{K'}{1 + K'}\right)$$

Figure 4.3.   Resolution equation.

For our purposes, the 5σ width is more useful for determining V_w. It is determined by drawing tangents to both sides of the peak and measuring the distance between where these intersect the baseline. Using this definition of peak width, the calculation of N equals 16 times the square of V_B/V_w. Different peaks in a mixture will give different efficiencies.

All of these are combined in the resolution equation (R_s), which predicts how each factor will affect the separation. The derivation of the equation is not important to our work, but can be found in the Snyder and Kirkland reference in Appendix E. In practice, the values used for the factors are empirically derived from chromatograms. For most uses, fairly crude measurements are sufficient, but care should be taken with peak widths in calculating efficiencies.

In general, k's range from 1 to 8 for analytical and 4–12 for preparative separations. Alphas range from 1 to 2; at $\alpha = 1$ peaks completely overlap, much above $\alpha = 2$ and the separation can be made in a separatory funnel. For N, values may range from hundreds (poor resolutions) to tens of thousands (good resolution).

The resolution equation shown in Figure 4.3 can point directions for starting separation scouting. Note that N is present as a square root term; large changes produce a small effect on resolution. k' is present in a convergent term. At low k's, a one unit change in k' produces a relatively large effect. At high k's a one unit change has little effect. This is why changes in k' above 8 have little effect except to lengthen the time of the run. Changes in α produce the greatest changes in resolution, but the exact effect that a given change in experimental conditions will have on the α value of a set of peaks is often difficult to predict. An α change in methods development is often saved as the court of last resort. It usually must be followed by further k' or N modifications.

Therefore, as we begin to develop a separation initially we will ignore column efficiency, knowing that we can use it later to produce small changes. We will make changes in retention until we reach high values of k'. Then, if we still have not achieved our separation, we will do something to change α.

Now let us look at the variable controlling the various factors in the equation. We will return to the resolution equation when we get into column diagnostics and healing (Chapter 6) and, again, in scouting and methods development (Chapter 12).

4.1.2 Efficiency Factor

The efficiency factor, N (Fig. 4.4), measures peak sharpness. The sharper the peak the better the separation and the higher the efficiency of the column and the system.

$$N = 16 \left(V_X \Big/ W_X \right)^2$$

Figure 4.4. Efficiency factor equation.

It is important, first, to realize that efficiency is not solely a function of the column. Bad extracolumn parameters, such as detector cell volume or tubing diameters, can make the best column look terrible. Second, efficiency measurements are very poor ways of comparing or purchasing columns unless all other parameters are constant. Many columns are bought and sold because they have a "higher plate count" than someone else's column. The efficiency calculations could have been made with different equations, on different compounds, on different machines, or at different flow rates, all of which will have a profound effect on efficiency. The only valid use of plate counts that I have found is in column comparisons where all other variables are equal, or in following column aging over a period of days or months.

Let us look at an efficiency measurement. Efficiency, N, is usually reported in plates, a dimensionless term that is a throwback to the days of open column, flooded plate distillations. The more plates in the distillation column, the more equilibrations occurred and the better the separation that was produced. In an HPLC column, the larger the plate count, the sharper the peaks are and the lower the amount of overlap that occurs between them.

For accurate measurement, it is important to spread the peak without changing variables affecting N. Increasing the chart speed to two to five times normal run speed will usually do this, but remember to correct V_B for the increase. Early eluting peaks with a k' of 1–3 should show a plate count between 6000 and 10,000 for a 10-μm packing in a typical 25 cm $\times$ 4 mm column.

Variables affecting changes in N have a square root effect on resolutions. Some are beyond the chromatographer's control, such as particle homogeneity, particle shape, and how well the column was packed. Particle size is a Gaussian distribution around the stated diameter. Different processing produces different distribution curves. Early packing produced by grinding and screening yielded very irregular-shaped particles with efficiencies lower than modern spherical particles. Packing is still very much of an art. Wall and bed voids act as turbulent mixers and are present to some degree in all columns. Spherical packing and high-pressure packing pumps seem to greatly reduce voiding and increase column life. Other variables, such as particle diameter and column length, are user-selected when the column is purchased. General analytical plates/meter for differing packages are shown in Table 4.1. These values are offered simply as a guide. Values of theoretical plates and optimum flow rate will vary for spherical packings and columns from different manufacturers. Column backpressures increase with smaller particle size and higher flow rates.

Column length is usually optimized around a tradeoff between efficiency and run time. Doubling the column length increases backpressure and run

Table 4.1 Relationship of Efficiency to Flow Rate

	Efficiency Changes with Particle Size	
Packing Diameter (μm)	Plates/Meter	Flow Rate (ml/min)
10	30,000	1.0
5	50,000	1.5
3	100,000	2.5

times 2-fold while increasing efficiency only by 1.4-fold due to increased diffusion.

Finally, we have variables affecting efficiency that can be controlled at the time of the run. These are pump flow rate, extracolumn volumes in the instrument used, and the method of calculation. Flow rate is the major efficiency variable that I use during methods development. Generally, halving the flow rate will increase separation around 40%. I do much of my scouting at 2.0 ml/ min, knowing that I can improve separation by dropping to 1.0 ml/min. Plotting of efficiency versus flow rate shows that each diameter of packing has its own optimum flow rate. Efficiency decreases at higher flow rates. In the microparticulate packings, large packing diameters show a more rapid loss of efficiency with increasing flow rate than do smaller packings.

Decreasing extracolumn volumes is critical to HPLC success. The most important volumes are those immediately adjacent to the column: zero dead volume endfittings, column inlet and outlet tubing diameters, and detector cell volumes. From the time the sample enters the injector until it exits the detector, nothing must add extra mixing space. Tubing from injector to column must be 0.010 in. for 5- and 10-μm packings with tubing lengths no more than 4–6 in. for the 5-μm packing. For 3-μm packing use 0.007-in. tubing about 3 in. long or less. Zero dead volume endcaps and connectors must be prepared correctly, so that tubing butts firmly against the fitting. We covered the preparation of compression fittings in Chapter 3, but if you find that efficiency drops after you change a fitting, check the dead volume fit. For detector cells, the rule is 8–12 μl cell volume; anything larger acts increasingly as a mixer for your already separated bands.

Tubing volumes outside the critical injector-to-detector range are important only if you are doing recycling or collecting samples. Pump-to-injector tubing is generally 0.020 in.; vents, flush valve, and so on may use 0.0040 in. Be sure you know what these look like and do not confuse them with injector tubing. In telling tubing apart, 0.02 and 0.01 in. are the most difficult. If you have to look twice to make sure there really is a hole, it is probably 0.01 in. If you are in doubt, put them next to each other. By comparison, 0.04-in. tubing looks like a sewer pipe.

There are many methods used to calculate efficiency. All methods give the same results with ideal, Gaussian peaks. Real chromatography peaks tend to tail on the back side (away from the injection mark on the chromatogram). When column problems occur they tend to show up as increased tailing. Calculation methods that use a peak width high on the peak miss these changes and give artificially high efficiencies. The 5σ method described above is excellent for detecting the early appearance of tailing. If you are planning on using a calculation using half-peak width make sure there is some method of measuring and correcting for peak asymmetry.

4.1.3 Retention Factor

The retention factor, k'_B (Fig. 4.5), also called the capacity factor, is the usual starting point for methods development. The retention factor, as its name implies, is basically a measure of how long each compound stays on the column; V_0 used to determine k' is usually only roughly measured; k' is a simply measured as a multiple of the V_0 distance.

The major usable variable controlling k' is solvent polarity. While temperature and column polarity also affect retention times, they do not show the same direct, linear relationship for all peaks and are classed under the separation factor (α).

Increasing the polarity difference between the stationary and mobile phases increases the retention of compounds with polarities more like the column. Compounds stick tighter and peaks will broaden through diffusion. Decreasing the polarity difference makes things come off faster. Peaks will be less resolved and sharper.

For example, for a polar silica column equilibrated with a mobile phase of methylene chloride in hexane (nonpolar), you would add more hexane to increase the polarity difference and the k' of relatively polar components. Adding methylene chloride, the more polar of the two solvents, would decrease the polarity difference and the k's, causing all components to wash off faster. With k' changes, peak position changes are proportional and in the same direction. The order of resolved peaks will remain the same; unresolved peaks should begin to pull apart.

If in our model system we had used 80% methylene chloride/hexane and the red peak had partially overlapped the back of the blue peak, we would attempt to resolve it by reequilibrating in 40% methylene chloride/hexane and reinjecting. We could expect that we should see two well-resolved peaks; if not, we could go to a 20% mixture. More than likely we would have overshot on the first change and would have to finetune back toward the 80% mixture. Simply

$$K = \left(V_x - V_o \right) \Big/ V_o$$

Figure 4.5. Retention factor equation.

by modifying the solvent polarity we are able to increase or decrease k' and contract or spread our separation. This k' development is our usual starting point in methods development.

So far I have referred only to "normal-phase" separations on polar columns. However, around 80% of the separations in the literature are made on "reversed-phase" columns. To understand these terms we need a little history.

Early "high-pressure" packings were crosslinked ion-exchange resins and polymeric size separation, gel permeation packings. The first high-pressure columns for partition separations were packed with the same material as is used in open columns or for TLC plates. This was 35 to 60-μm-diameter silica, a very polar packing material. To achieve separation, nonpolar solvents were used. These solvents were flammable, volatile, toxic, or expensive. After a few years, someone decided to coat the silica with nonpolar compounds similar to those used in GLC columns so that polar solvents, such as water, could be used for elution. The problem with these coatings was that they tended to wash out with the mobile phase, bleed into the detector, and contaminate the collected sample.

This was overcome by chemically bonding the coating to silica, leading to the first "abnormal" packing materials. Because these packings could be run in aqueous solvents and did not require the careful drying and handling of the normal-phase columns, they quickly became very popular. Since no one wanted to admit to being a "abnormal" chromatographer, when they reached the majority they quickly renamed themselves "reversed-phase" chromatographers.

The first of the really successful coatings was a long-chain, saturated hydrocarbon with 18 carbons. These octadecyl- (ODS), RP$_{18}$, or C$_{18}$ columns are still the most commonly used HPLC columns, primarily because of the versatility they have shown. Other packing materials have appeared with shorter or longer side chains, and with a variety of functional groups on the side chains, greatly extended the possible separations that can be achieved with HPLC.

Retention changes work exactly the same with reverse-phase as with normal-phase columns. Increasing the polarity difference between column and mobile phase increases the k's of the components. However, since the column is nonpolar, we now must add more of the polar solvent to make compounds stick tighter. On our reverse-phase column, our dye mixture would also elute in opposite order; the more polar red dye would have less affinity for the nonpolar column and would elute before the nonpolar blue dye. By controlling the column nature, you control the elution order. Figure 4.6 illustrates the effect of solvent polarity changes on a separation.

As we mentioned earlier, there is a limit to the usefulness of k' changes. Because it is a convergent term in the resolution equation, the larger the value of k', the less the effect a polarity change has on R_s. Beyond $k' = 8$–10, changing k' has only a negligible effect, except on run time. At this point, the next step is to change resolution, R_s, by using the separation factor, α.

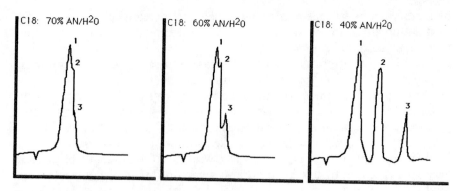

Figure 4.6. Effect of polarity changes.

4.1.4 Separation (Chemistry) Factor

The separation factor, α (Fig. 4.7), is calculated by dividing the k's for the two peaks under question. It measures the separation betweeen the two peak centers. Components with an α of 1.0 overlap completely; beyond 2.0 compounds can be separated by separatory funnel. Large αs are needed in HPLC only for preparative runs.

When we change retention with solvent polarity, all peaks show an equivalent shifting in the same direction. A variable producing an α change causes peaks to shift, but individual peaks exhibit different amounts of shift, both in size and direction. Thus, k' changes spread separations already present; with α new separations are created. With an α change relative peak positions can even reverse.

Temperature is the first of the variables affecting separation. Increased temperature decreases retention time on the column, sharpens peaks, and produces the change in relative peak retentions typical of an α effect. At first this appears to be the ideal variable, similar to temperature programming for GLC. However, temperature has some drawbacks.

First, temperature is generally limited to an effective range of 20–60° C by solvent vapor pressures. It can produce chemical changes catalyzed by contact with the acidic silica surface in some compounds being separated. Even more important is the effect temperature has on the column packing. Bonded phase columns are prepared by chemically bonding an alkyl chlorosilane to the oxygen on the silica surface. This process can be reversed by hydrolysis, especially under acidic conditions, leading to bonded phase bleeding and column perfor-

$$\alpha = \left(V_X - V_O\right) \Big/ \left(V_Y - V_O\right)$$

Figure 4.7. Separation factor equation.

mance changes. Heat accelerates the process. If you are getting only 3 months of life from your columns this might not be an important consideration. However, one of our goals is to show you how to extend column life.

The separation factor is also referred to as the chemistry factor. It can be modified by changes in the chemistry of the components that make up the chromatographic system: column, solvent, and sample. Changing the column surface chemistry from the very nonpolar C_{18} to C_8 obviously increases the column polarity as the compounds are drawn closer to the silica surface. We would predict that nonpolar compounds would elute faster, and so they do. However, observation of the peaks shows peak shifting typical of an α variable. If we substitute a phenylethyl group for an octyl group, we maintain about the same polarity, but now we see dramatic changes in selectivity. The so called phenyl column has an affinity for aromatics and double bonds. It will separate fatty acids on the basis of the number of double bonds as well as chain length. Octyl columns separate only on chain length differences.

The most common variable used to control α is the "stronger" solvent in the mobile phase. The stronger solvent is the mobile phase component most like the column in polarity. Changing the chemical nature of this stronger solvent will produce shifts in the relative peak positions. For instance, if we are unable to achieve the desired separation on a C_{18} column using acetonitrile in water, we can produce an α effect by shifting to methanol in water. An opposite effect occurs on switching to tetrahydrofuran in water (see Fig. 4.8).

This is true even if we adjust the polarity of the new mixtures to match that of the previous mobile phase. We can produce other α changes by adding *mobile phase modifiers* to our solvents. Buffers, chelators, ion pairing reagents, and organic modifiers can all be used to change or finetune the separation. We will cover all of these in detail in Chapter 7.

The final α modifier, preparing derivitives of a mixture, is our court of last resort. If two compounds cannot be separated by changing N, k', or the chemistry of the column or mobile phase, then changing their chemical nature by making derivatives should lead to compounds that can be separated. We use

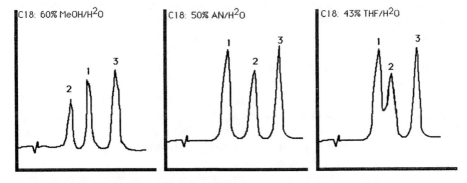

Figure 4.8. Effect of "stronger" solvent changes.

this only as a last option separation technique. Usually we can separate most compounds directly. Derivatives are more commonly used in HPLC to change a mixture's solubility or to produce compounds with strong extinction coefficients to increase detection sensitivity.

4.2 Ion-Exchange Chromatography

So far, we have dealt only with partition chromatography in which compounds equilibrate between the mobile phase and the column based on differences in their polarity. Ion-exchange chromatography uses the type and degree of ionization of the column and compounds to achieve a separation. Here opposites rather than likes attract; compounds with charges opposite to that on the column are attracted and held by the column. Elution is achieved by competitive displacement; an excess of an ion with the same charge as the bound compound pushes it off the column. The tighter the ionic bonding to the column, the longer the compound stays on the column.

Ion-exchange columns are made of two backbone materials: silica, like the reverse-phase columns, and heavily crosslinked, organic polymers. Bound to these are organic bonded phases containing functional groups that either have permanent ionic charges or in which ionic charges can be induced with pH changes.

Two warnings about using polymeric columns. Early polymeric columns for ion exchange would not tolerate much pressure or organic solvents. Recent columns are more heavily crosslinked and show more pressure tolerance, but be sure to check the manufacturer's column shipping notes for use limitations. Few will tolerate pressures above 2000 psi without collapsing. Some organic solvents can cause the column bed to swell or shrink on changing solvents, which can lead to bed collapse or voiding.

Charged functional groups, which give these columns their separating characters, are of two types: anionic and cationic. Anionic packing materials have an affinity for anions (negatively charged ions) and have positively charged functional groups on their surfaces, usually organic amines. Cationic packings attract cations (positive charges) with bound negative functionalities, usually organic acids and sulfonates. Cationic and anionic columns can both be subdivided into either strong or weak types. Strong columns have functional groups that possess either permanent charges (i.e., quarternary amines) or that have charges present through the full pH range used for HPLC (i.e., sulfonic acids). Weak columns have function groups with inducible charges. At one pH they are uncharged and at a different pH they are charged. Examples are organic acids, which are uncharged at pH 2.0, but form cations at pH 6.5, and organic primary amines, which are positively charged below pH 8.0, but exist in the free amine form above pH 12.

Let us examine a silica-based cationic (sulfonate) ion-exchange separation (Fig. 4.9). The column is equilibrated in 50 mM sodium acetate. A solution of

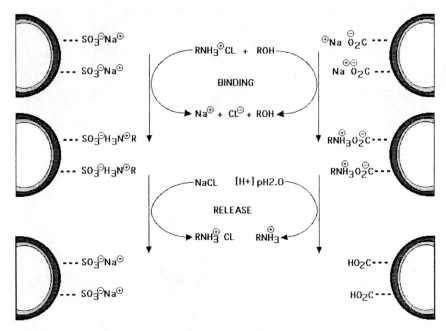

Figure 4.9. Cationic-exchange separation model.

amines and an alcohol in the mobile phase is injected. The same mobile phase, or one containing increased amounts of sodium acetate, is used to elute fractions.

The alcohol will come off in the void volume of the column since it has no attraction to the column. The amines will be retained, because at the pH of the acetate solution they are protonated and have a positive charge. As more mobile phase passes through the column, its sodium ions begin to compete for the sulfonate sites with the bound amines. Through a mass effect the amines are displaced down the column, until, finally, they elute into the detector. The amine that has the strongest charge and binds the tightest is eluted last.

4.3 Size-Exclusion Chromatography

The first commercial HPLC system was sold to do gel permeation or size-separation chromatography (GPC). It is the simplest type of chromatography, theoretically involving a pure mechanical separation based on molecular size.

The column packing material surface is visualized as beads containing tapered pits or pores. As the mobile phase sweeps the injection past these pits, the dissolved compounds penetrate, if their largest diameter (Stokes radius) is small enough to fit (Fig. 4.10). If not, they wash down the column with the

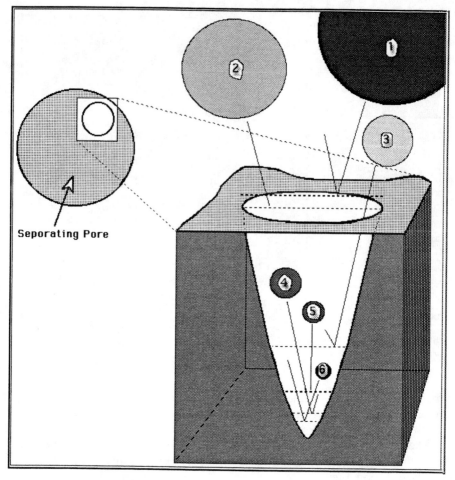

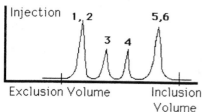

Figure 4.10.   Size-separation model.

injection front and elute as a peak at the column void volume, which is called the *exclusion volume.*

Returning the compounds that entered the pit, we find that large particles cannot penetrate as deeply down the pore as can smaller compounds. The smaller the diameter the deeper the penetration, and the longer the compound takes to elute. The largest compounds wash out more quickly, follow a longer path, and elute later than the totally excluded compounds. Traveling down the column, these resolving compounds wash in and out of many pores, magnifying the resolution achieved by differences in the path lengths they follow. Finally we reach a point where all compounds of a certain diameter or smaller reach the pore bottom, wash out, and elute in a single peak. This is referred to as the *inclusion volume*. If the exclusion volume is found at V_0, the inclusion volume appears at approximately $2V_0$.

From this we can see we have three types of peaks: (1) the exclusion peak, containing all molecules of a certain size or larger; (2) resolved peaks of intermediate diameter; and (3) the inclusion peak containing all compounds of a given diameter and smaller. In a crude mixture of compounds, we are forced to suspect that both the exclusion and inclusion peaks contain multiple components.

Just as it is possible to prepare a column with a single pore size, it is possible to prepare columns with differing pore sizes. Each would have its own particular ratio of exclusion/inclusion diameters. A column bank can be used to separate a mixture with a wide range of compound sizes. Columns of varying exclusion/inclusion limits can be connected, with the smallest exclusion limit column first in the series. If the columns are selected so the first's exclusion limit overlaps the second's inclusion limit and so forth, the column bank produced has the first column's inclusion limit and the last column's exclusion limit. Again, remember the pressure problem when stacking columns; pressure increases proportionally to the number of columns. You may have to run very slowly if you are using pressure-fragile columns.

GPC columns are referred to as molecular weight columns, but actually they separate molecules according to their largest dimension. True molecular weight measurements would be independent of shape. As long as we work with simple, spherical compounds there is a direct relation between exclusion volume and molecular weight within the resolved range. Columns can be calibrated with standards of known molecular weight and used for molecular weight determinations. These measurements break down at higher molecular weights with compounds of nonspherical shapes (i.e., proteins), which change shape and apparent size with changes in the mobile phase. Solvent conditions that force all molecules into long, rigid shapes aid in molecular weight determinations (i.e., 0.1% SDS for protein *molecular weights*).

Size-separation columns are available with both silica and heavily cross-linked organic polymer backbones. The polymer columns show the same pressure and solvent fragility described for ion-exchange columns. Silica size columns must be protected from pH changes like partition columns, which have a pH between 2.5 and 7.5.

4.4 Affinity Chromatography

Much less common than partition, ion exchange, and size columns, affinity columns are of growing interest in the HPLC purification of proteins because of their very high specificity. Bound to the surface of the affinity packing, sometimes through a 6 carbon spacer, is a molecule with a target site or recognizer. This forms a tight complex with one, and usually only one, site on the compound to be purified. The analogy used in affinity separations is one of the *lock and key*. The target site on the compound to be separated is the key and the recognizer on the affinity packing is the lock. When a solution containing the target compound is placed on the affinity column, only that material is held up. Everything else comes out in the breakthrough. The target compound can then be removed with a change in pH, with high salt, or eluted with a molecule similar to the recognizer compound.

In practice, affinity column recognition specificity is never as complete as described in theory. Usually a range or class of compounds can be attracted and retained. The recognizer must be bound to the column for each target compound and after that point the column must be dedicated for that purpose. Usually there is no possibility of removing the recognizer and reusing the column for a different separation.

The biggest attraction of this type of column is that often it is able to achieve nearly a total purification of the target from a very complex mixture in a single pass down the column. Like the ion-exchange column, this type of separation benefits in a preparative mode, from broad, short columns with a large surface area. Its weakness lies in the difficulty of finding and binding the specific recognizer for our target, and in developing optimum eluting conditions.

5

Column Preparation

The power of HPLC is rooted in the variety of separations that can be achieved with little, if any, sample preparation. HPLC columns are often described as the "heart of the separation." Controlling a separation means understanding and controlling the chemistry and physics going on inside of the column. To do so it is necessary to understand how packings are prepared and how columns are packed. This will lead us to methods to keep columns up and running, to an understanding of when to select a given column, and to techniques for getting the most from that column.

5.1 Column Variations

The first packing materials used in a HPLC were beads of organic gel permeation resins used for size separations. These were commercially available resins and no attempt was made to optimize them for high pressure, except to select for a high degree of crosslinking to prevent crushing.

A year later silica, fully porous 35- to 60-μm-diameter beads, was slurry packed in a tube and used for separation. This was the same material that had been used for open column or thin-layer chromatography. The only gain over these earlier techniques was in separation time. Almost immediately, research was begun to optimize the packing to improve the separation.

It was soon found that a large amount of band spreading occurred in this material because of the variety of particle diameters, leading to wide variation in paths a compound could follow going through the packing. Screening of the packing allows separation of a fraction with an approximate diameter of 35 μm. This *porous* packing gave a better separation and because of its high porosity and corresponding high load capacity is still used today for preparative separations.

The next advance came with the discovery that interparticle path variations were contributing to band spreading. With large, fully porous materials, compounds could follow a separation path either through the diameter or barely skimming the particle surface. It was like having a mixture of particles with diameters from 35 μm on down. The more uniform the path followed, the higher the expected efficiency of the separation would be. To achieve this, a crust of porous silica was coated on the outside of a solid, glassy core forming a *pellicular* packing. This was the first of the true analytical packings. Its 35 μm diameter made it easy to handle and pack, its uniform separation path gave it good efficiency, but it had very poor loading characteristics for preparative work. This packing is still used for packing guard columns to protect 10-μm analytical columns from contamination.

The next major step was to the *microporous* analytical packings. These fully porous 10-μm packings were prepared by grinding and selectively screening the 35–60 μm fully porous materials. Although irregular in shape, they had very high efficiency and very good load characteristics. They suffer from two basic problems: high back pressure and fines. Because of small diameters, they pack very tightly and provide considerable flow resistance. Modern HPLC pumps capable of 6000–10,000 psi appeared in response to these packings. The small size and irregular shapes also made it difficult to pack these materials without trapping solvent in pockets in the bed and along the wall. The voids formed led to efficiency loss by acting as turbulent remixers and premature death of the column by channeling. Fines were carefully washed out of these packings, but reappeared, plugging the outlet filter during use. The packing suffered from microfractures and released ground off fines as the bed suffered movement during pressure changes.

The most recent improvement has been the fully *spherical* microporous packings. Under an electron microscope, these packings appear as true spheres, usually either 3 or 5 μm in diameter. Not all particles in a batch have exactly the listed diameter; they show a distribution around that size. Each spherical particle shows a single uniform diameter while the irregular micropacking shows a major and a minor axis. Irregulars also show fissures and grooves while the spheres are featureless snowballs. The spherical packings give a more uniform bed, a slightly higher backpressure, and have little tendency to void unless solvent etched. They are the packings of choice for new methods development.

Little was known about the process used to prepare these packings until one manufacturer gave a clue in a technical brochure. Molten silica is cooled at a controlled rate in a polymerizing organic matrix. The plastic formed is then sintered off leaving the microporous spheres behind. A delicate process is needed to control diameter, shape, and porosity during preparation.

Normal-phase silica packing requires only drying at a uniform temperature to be ready for packing. At 250° C the fully hydrated silica is produced, while at 300° C water is lost between adjacent silica molecules forming the anhydride form normally packed in normal-phase columns.

The various bonded-phase columns require a bit more processing. The first step, *silylation,* involves reacting fully hydroxylated silica with a chlorodimethylalkylsilane and heating to drive off HCl. Variations in the chain length and functional groups on the alkyl group produce a wide variety of bonded-phase columns. If we stop here, we would have a column that gives good sep-

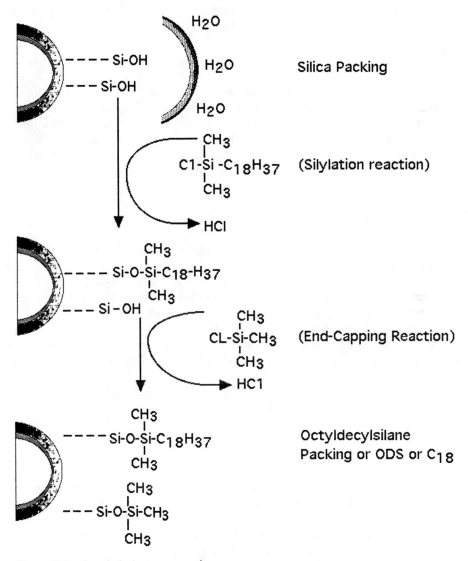

Figure 5.1. Bonded-phase preparation.

arations for acidic and neutral compounds, but that gives very poor, tailing separations of amines and bases. Steric hindrance prevents complete bonding of all the free silanol sites; about 10% of the available sites are still free (Fig. 5.1).

The next step is a process called *endcapping*. This involves bonding of the remaining silanols with a smaller compound, chlorotrimethylsilane. After this treatment, free silanols are $<1\%$ and the column can be used for amine separations. The process by which these bonded groups are attached is reversible in the presence of water at either low or high pH. In the past, dichloro- and chloroalkylsilanes have been used for silylations, producing multilinked, crosslinked, or polymeric coatings. Controlling the degree of silica hydration also controls the amount of coating that attaches. When you consider the different columns that can be produced by controlling hydration, bonding agent, coating levels, and end capping, it is not hard to understand the variations in C_{18} columns coming from different manufacturers. A recent paper in *LC/GC Magazine* announces the preparation of a C_{18} column with much better resistance to temperature and pH changes. Not endcapped, it uses coatings prepared by reacting chlorodiisopropyloctyldecylsilane. The diiopropyl groups shield the unsilyated $-OH$ allowing amines to be run without tailing and protecting the $Si-O-SI$ linkage from acid-catalyzed hydrolysis. Other C_{18} materials have been prepared on polymeric, zirconium, and neutral alumina base supports.

5.2 Packing Materials and Hardware

Column packing is as much art as it is science. Even the professionals in the field cannot routinely prepare columns that will give exactly the same plate count column to column. They quality control the columns with a set of standards, and columns that deviate by more than a set amount are either dumped or repacked (not cost effective), or are sold as specialty columns. Columns that exceed QC specifications are almost as bad as poor efficiency columns; methods developed for "standard" columns will not work on these "super" columns. To use them you have to repeat at least part of your methods development, which means lost time.

You may find that for teaching purposes packing your own columns may be cost effective. An analytical column holds about 3 g of packing and the column itself can be dumped, cleaned, and reused. However, I would recommend not packing your own research columns. I will show you ways of extending the lifetimes of your columns enough that commercial columns should be cost effective.

Column packing is in theory very simple, but proves most difficult in practice. Packing material is mixed in a viscous solvent and driven with a high-pressure pump into a column. Solvent passes out of the column through the end-fitting filter, while solids build up and pack down on the frit. As the column fills with a packed bed, back pressure increases until the column is filled. Once the column is filled, the slurry reservoir is removed, excess packing is scraped off, and the inlet frit and endcap are attached. Care must be taken to ensure that no packing material is left in the threads of the endcap. Silica is an excellent abrasive and will score the stainless steel, leading to leaking.

A number of commercial column packing apparatuses are available. One type, the ascending type, is a stirred can that pumps slurry upward and then down into the column. The descending type is simply a slurry reservoir that attaches in place of the inlet endcap and frit and is equipped with a pump connection at the top (Fig. 5.2). Manufacturers use 20,000 psi pumps to drive

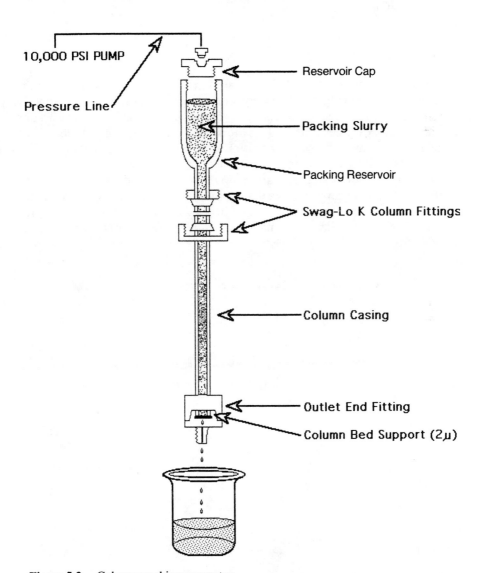

Figure 5.2.   Column packing apparatus.

slurry into the column, but most laboratory packing apparatuses rely on pumps that reach a maximum of only 6000–10,000 psi. The pumps are run fully open until the pressure stabilizes.

Once packed, the column needs to be checked for efficiency using column standards. We have discussed storable column standards, but if you will be using the column with amines, it might be a good idea to add fresh amine of known running characteristic to the mixture. Amine tailing is a very good check for tailing or endcapping problems, but amines air oxidize and are not stable for long storage.

5.3 Column Selection

Selecting a column for an HPLC separation is a matter of asking yourself a series of questions (Fig. 5.3). You must first determine how much material you wish to separate in a single injection (preparative vs. semipreparative vs. analytical). The next question involves the separation mode to be employed (size exclusion vs. ion exchange vs. partition). Finally, there is the question of solubility controlling solvent and column selection in all modes.

If your selection is *size* separation, do the molecules you are trying to separate vary by size or molecular weight? If the differences are in size, how large is the range of differences and how close in size is each pair of compounds that must be separated? Size columns are rated by inclusion/exclusion range and

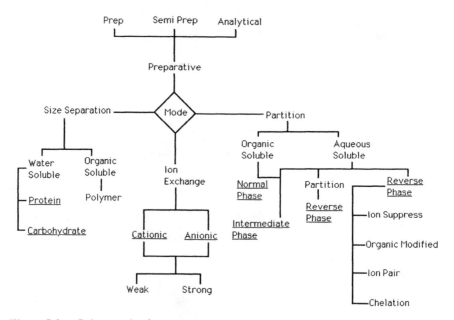

Figure 5.3.   Column selection.

the separating molecules must fall in this range to be resolved. Generally, it is difficult to separate two compounds that differ by less than 100% in molecular weight.

Do your compounds differ by *charge* or have a charge that can be influenced by adjacent substituents? It is very easy to separate a charged molecule from an uncharged molecule or a molecule of a differing charge on an ion-exchange column. It is simply a matter of selecting a column that has a charge opposite the compound in question (anionic vs. cationic). With compounds that have the same type of charge, the separation is made based on the way the electron density of the charge is modified by steric difference and functional groups near the charge site. To get resolution, we often need to control the column charge strength and the pH, or use eluting salt gradients to remove selectively the components of the mixture.

Are the primary differences in *polarity?* Partition columns are available that vary in polarity from nonpolar (octyldecyl), through intermediate polarity (octyl and cyanopropyl), to polar (silica). Some columns have similar polarities, but differ in their specificity. C_{18} and the "phenyl" column have similar polarities, but C_{18} separates on carbon chain length, while phenyl separates fatty acids on both carbon number and number of double bonds. Phenyl columns also resolve aromatic compounds from aliphatics of similar carbon number. In another example of similar polarities, C_8 is a carbon number separator while cyanopropyl selects for functional groups.

Assuming we have selected the proper mode of chromatography, will the mixture dissolve in the mobile phase? Ion-exchange columns must be run in polar charged solvents. Size-separation columns are not, in theory, affected by solvent polarity, and size columns for use in polar and nonpolar columns are available. In partition chromatography, we have nonpolar columns that can be run in polar or aqueous solvents, and polar columns that are run only in anhydrous, nonpolar solvents. Intermediate columns such as cyanopropyl or diol can be run in either nonpolar solvent, although often with differing specificity. An amino column (actually a propylamino) in methylene chloride/hexane acts like a less polar silica column, but in acetonitrile/water can be used to separate carbohydrates by carbon numbers.

I tend to think of the columns as being a continuous series of increasing polarity from C_{18} to silica: C_{18}, phenyl, C_8, cyano, C_3, diol, amino, and silica (Fig. 5.4). Under that I have solvents used to prepare mobile phases in a similar order of polarity from hexane under C_{18} to water under silica: water, MeOH, *i*-PrOH, acetonitrile, THF, methylene chloride, chloroform, benzene, and hexane. The cyano column and THF are about equivalent polarity. In setting up a separation system we cross over; nonpolar columns require polar mobile phase and vice versa to achieve a polarity difference.

To make a separation, I look at the polarity of the compound I want (X) and its impurity (Y). Like attracts like. Let us assume the compound is more non-polar than its impurity. On a C_{18} column, the nonpolar compounds stick tightest to the nonpolar column and elute last; the impurity comes off first.

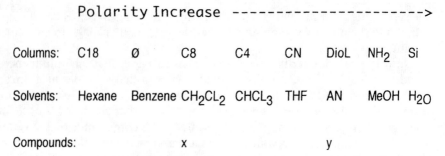

Figure 5.4. Polarity trends—columns and solvents.

Running the same separation on a silica column in a nonpolar solvent, we should expect a reversal. The polar impurity sticks to the polar column, while the nonpolar compound washes out first in the nonpolar solvent. By thinking about the polarities involved in the separation, we can control the separation.

We are not limited to a single column type or chromatography mode in our attempt to achieve a separation. We can use a technique called *sequential analysis* (Fig. 5.5). For example, we can make a size separation, then take a size fraction and do a partition separation. This is commonly used in separating a complex biological mixture where a single separation mode would be overwhelmed. Separation on first a C_{18} and then a silica column is often used to confirm purity of a compound. If it passes separation on two different types of columns, it is a fairly good bet that it is pure. Even better is to confirm identity by using two differing separation modes, for instance partition and ion exchange.

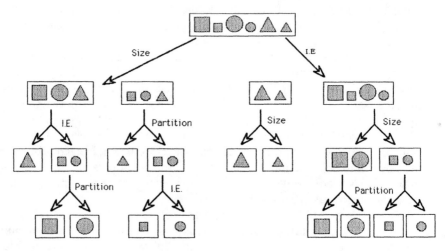

Figure 5.5. Sequential analysis.

There is one more type of column you might want to select for your separation, the *protective column.* This column is designed to protect the column bed, and, as such, it is a sacrificial column. There are two basic types of in-line protective columns: the guard column and the saturation column (Fig. 5.6). The disposable cartridge column or sample preparation column (SPE) used in sample preparation also serves a role in column protection and will be covered in Chapter 12.

Guard columns are a minianalytical column pressure packed with the same material used in the analytical column. It is connected in the path from the injector and collects anything that normally would be deposited on the main column. It must be cleaned or replaced periodically because contamination will eventually bleed through. Since guard columns usually are only about 1–3 cm in length, they can be inverted and backwashed without causing them to void. Please do not wash the guard column down the main column. Disconnect and reverse it, and reconnect using the pump to wash a strong solvent through it into the beaker. This may seem obvious, but I had to troubleshoot a persistent detector problem that turned out to be caused by washing a guard column into a main column.

Since the guard column is placed in the injector/column path, it does contribute to the separation. Methods development should be completed with the guard column in place. The increased separating length usually overcomes the effect of extra tubing as long as the connecting tubes are kept as short and as fine as possible. The wrong diameter tubing can really mess up a separation. Changing guard columns in the middle of a series of runs generally has little effect on the separation. However, it is usually a good idea to follow the change with a standard QA run as a check.

The other type of protective, in-line column is the saturation column. This column is used when operating conditions tend to dissolve the main column bed (i.e., high pH, high temperature, etc.). In theory, the packing in the satu-

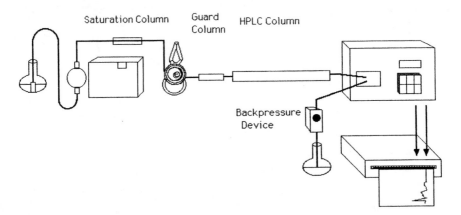

Figure 5.6.   Protective columns.

ration column dissolves first, protecting the main packing. As long as the same type of bonded phase is used in the precolumn, the column running characters do not seem to change. Using this technique, a customer ran taurine separations at pH 12 for a year on the same analytical column without creating pits or dissolving packing. Care must be taken that the saturation column does not break through; erosion of the main column will begin immediately if this happens. A guard column will serve as a saturation column, but is not recommended, since the precolumn bed is consumed and band spreading will occur.

Usually the saturation column is placed in the flow from pump to injector. At this location the column to be used can be slurry packed with large diameter packing with no regard given to packing efficiency. I have even seen columns dry packed with tamping, wetted with solvent, and placed in line as a saturation column. I am not entirely satisfied with the explanations as to why this technique works. I offer it to you as a tool that others have used to produce separations at pH high enough to separate many amines in their free amine form. Silica appears as a solid out of evaporating fractions and, occasionally, coats out on detector windows. I would recommend using this technique only when other methods have failed.

6

Column Aging, Diagnosis, and Healing

HPLC columns have a reputation of being fragile things that have only a limited lifetime and, therefore, are expensive to buy and maintain. Much of this reputation is undeserved, and in this chapter we will explore the aging of columns, the symptoms of aging, and methods of regenerating columns and extending their operating life. The typical new chromatographer gets about 3 months of life from a column; an experienced operator gets about 9 months. I hope to help you extend column life to 1–2 years.

I know this is possible from a bonded-phase column because I had a customer who averaged this on his columns. He ran a clinical laboratory and rotated C_{18} columns through a series of four separations, each less demanding than the one before it. When the column failed on separation 1, it was washed and reequilibrated for a less demanding separation 2, and so on.

Over the years, I have collected hints, ideas, and tips that were not available to him, allowing us to get the same performance from a single column without rotation. The key to treating column problems is to know when problems are occurring, and to catch them as early as possible and treat them. The main tool for early detection of problems is the column QA with standards described earlier and illustrated in Figure 6.1.

There are five basic types of "killers" of column efficiency. Killer 1 is effects that remove the bonded phase. Killer 2 is effects that dissolve the column surface or the packing itself. Killer 3 is materials that bind to the column. Killer 4 is things that cause pressure increases. Killer 5 is column channeling. There are definite symptoms of each of these and either treatments or preventions for each type of killer (Fig. 6.2).

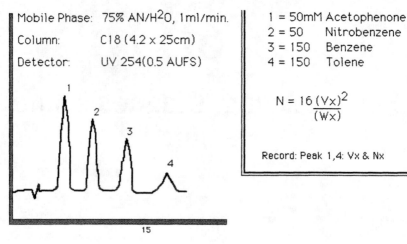

Figure 6.1. Column standards.

The best way to follow column changes is with column standard plate counts. We will use the four-standard mixture of acetophenone, nitrobenzene, benzene, and toluene described in the discussion on efficiency factors (Chapter 4). Our column will be a C_{18} reverse-phase column run in 70% acetonitrile/water at 254 nm. In an initial run, we obtain four peaks whose peak αs double between each pair. After we discuss reverse phase, we will see how these killers affect normal-phase columns.

      (1) <u>Bonded Phase Loss</u>
           a] low pH (2.0).
           b] high temperature (>ambient).

      (2) <u>End Voids</u> - dissolved column packing
           a] high pH (8.0).
           b] high salt (>200mM).

      (3) <u>Bound Compounds</u>
           a] non-polar organics.
           b] inorganic cations.
           c] ionic detergents/Proteins.

      (4) <u>Pressure Increases</u>
           a] Column: Inlet frit, outlet, bed.
           b] System.

      (5) <u>Center voids</u> - channeling.

Figure 6.2. Column killers.

6.1 Packing Degrading—Bonded-Phase Loss

Column degradation can be caused by too low pH or too high temperature. Columns should be operated in a pH range of 2.5–7.5 at ambient temperature. Below 2.0, bonded phase comes off and free silanols are formed, making the column more polar and increasing the cationic-exchange character of the surface. Our four-standard separation tends to collapse on the center of the four peaks (Fig. 6.3). More polar peaks retain longer, less polar peaks come off faster, and all peaks broaden and tail. Finally, we end with a single, very broad, badly tailing peak. This problem cannot be healed, only prevented. Attempts have been made to pass solutions of chlorotrimethylsilane down a degraded column, but they have not restored much activity. Control pH with buffers so that they do not fall below 2.0. There has been limited success using saturation columns where pHs below 2.0 must be used, but window coating with bonded phase is a common problem.

Elevated temperature can produce two different effects. Basically, it increases the solubility of the silica packing and, thereby, accelerates end void production like high pH. At low pH, it also accelerates bonded-phase removal, rapidly producing the four-standard peak effect seen at low pH. It is hard to believe that one manufacturer actually recommends temperature programming as a tool for gradient chromatography. It might work for silica columns in nonaqueous solvents, but I do not recommend it for bonded-phase columns unless you are planning on buying a lot of columns.

One special problem already alluded to is the oxidation of bonded phase containing amino groups, such as the propylamino group or DEAE packings. These amines will oxidize, turning yellow, brown, and eventually black just like a bottle of amines sitting on a shelf exposed to light and air.

I first encountered the problem while diagnosing a separation of carbohydrates using MeOH/water on an amino column. Columns were dying in 3 months or less. When we removed the endcap, the packing under the inlet frit was black. Dumping out the packing showed it to be darkened all the way down the column, even though a fresh column was white.

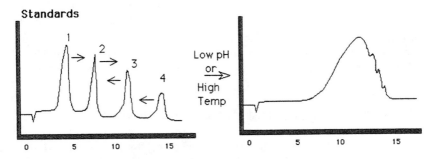

Figure 6.3. Effect of bonded-phase loss.

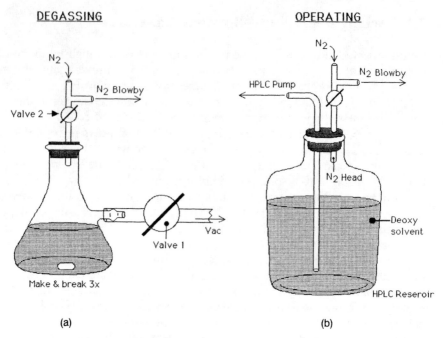

Figure 6.4.   Deoxygenation apparatus.

We solved the problem by giving the customer a new column and having him prepare and run only deoxygenated solvent in the solvent in his amino column using the apparatus shown in Figure 6.4. Solvent is vacuum degassed until large bubble formation stops, the vacuum valve is turned off, the nitrogen blowby turned on, and the inlet valve 2 slowly opened, allowing the vacuum to be broken with nitrogen (Fig. 6.4a). Vacuum is pulled and broken in this fashion three times. Next, a nitrogen purge is placed in the solvent reservoir bottle and oxygen is displaced. Deoxygenated solvent is pored down the side and the nitrogen blowby top is fitted to the top as in Figure 6.4b. The pump line is connected to the HPLC pump, the nitrogen blowby is turned on, the demand valve 2 is turned on and the pump is started, the system is purged up to the column, and the amino column is installed and equilibrated. At the end of the run the pump flow and the demand valve are turned off at the same time until they are needed. When not in use the amino column is stored in deoxygenated solvent. The customer in question got 14 months on his next amino column.

6.2 Dissolving Packing Material—End Voids

At high pH, above 8.0, silica begins to dissolve, forming an end void rapidly, even if protected with a bonded phase. To be safe it is best to keep pH below 7.5, unless the column is protected with a saturation column. The four-standard separation shows a progression of "rabbit ear" fine peak splitting, to a shoulder, to peak broadening on all four peaks (Fig. 6.5). If the column is opened at these three stages, increasing amounts of pitting and bed settling can be seen. At the rabbit ears stage, a fine pit directly in the bed center can be seen. By the shoulder stage, the pit has spread and the bed dips down on one side. By the time the shoulder disappears, enough of the bed has eroded so that a millimeter or so of packing is missing across the whole surface. Even though the peaks change appearance on pitting, the k's remain unchanged for the peaks. This allows us to distinguish between end voiding and organic contamination, which we will discuss later.

These end voids can be repaired; fresh packing material can be worked into a paste with mobile phase and pushed into the moistened pit with the flat side of a spatula. Overfill the column head, strike it off with a card, replace the end frit, and retighten the endcap. Be sure not to leave silica in the threads; wet with MeOH, use a Moore pipette to dry, then blow the treads clean. Reequilibrate the column with solvent and rerun the standards. If the pit is very deep, it may be necessary to repeat the repacking and pumping. Eventually, all peaks should be needle sharp again. Packing material is available from some manufacturers in small quantities. A gram should top up a lot of columns. Try to use the same size and type of material used originally in the column. If you cannot get 3 μm packing, use 5 μm packing from the same manufacturer. (The outlet end of "used" columns, discarded by chromatographers who do not know how to repair them, is an excellent source of clean packing.) If all else fails, pack them with glass beads of the same diameter.

Salt solutions with concentrations above 200 mM tend to erode column beds by increasing the ionization and, therefore, the solubility of the silica. The effect is similar to high pH end voiding and can be treated in the same way. One customer, who ran high salt gradient ion-exchange columns, solved his severe end voiding problem by amputation. He opened the endcap, cut the column below the end void with a tubing cutter, put a new column ferrule on with a crimper, and replaced the endcap. The new column was shorter and had less resolving power, but still worked. He continued cutting the column until it reached 3 cm, then used it for a guard column. I do not recall a customer using a saturation column to prevent salt erosion, but it should work. This effect occurs with nonhalide salts as well as halides; the extractor seems to be salt positive ion.

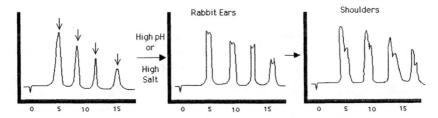

Figure 6.5. Effect of dissolved packing—end void.

6.3 Bound Material

The third type of column killer is material stuck to or coated on the surface of the packing that changes the column's running characteristics. The binding materials fall into three broad classes: organics, inorganic cations, and charged organics.

Uncharged, nonpolar organics sticking to the column tend to affect specifically the later running peaks in a separation. In the four-standard mixture run, it is the benzene and toluene peaks that broaden, shorten, and disappear (Fig. 6.6).

Contaminated water is a notorious source of this problem, and is the usual place to look for the culprit. One of the quirks of human nature is that people refuse to admit that their water could possibly be contaminated. I have seen triple distilled water that worked fine for enzyme reactions fail miserably for HPLC. I once spent 9 months convincing a friend that his PTH amino acid gradient separation was losing its last two peaks because of bad water. After washing the column and switching to HPLC grade water, the problem disappeared, never to return. Unextracted injection samples are the second source of organic contamination. If you find your baseline rising and falling when you are just pumping mobile phase through your column, there is probably nothing wrong with either your detector or the pump. When a baseline goes up and down, it almost always means that a peak has just come off the column, no matter how broad the peak or how close the peak maximum is to the original baseline. Garbage on the column eventually washes off. As it starts to come off, the baseline goes up. When it finally finishes, the baseline goes down.

I have had many people threaten to return whole HPLC systems because of "bad pumps" or "bad detectors," when the systems were suffering from dirty columns. It is always the detector first and then the pump that gets blamed. And, the poor service people who are hardware oriented, as they usually are, will make multiple trips without finding the problem. I encourage our service people to carry a C_{18} column, used only with standards, and a vial of four-standard mix in their service bag. The first thing they do is remove the customer's column and run the four-standards in their column. It is very

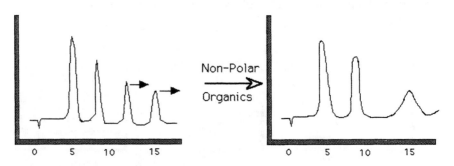

Figure 6.6.   Effect of bound nonpolar material.

embarrassing when the "detector" or "pump" problems go away, but it saves the company or the customer a lot of money.

The problem in this case is usually nonpolar organics (the polar organics do not stick to nonpolar columns, but wash through the column leading to an elevated baseline). Washing the column will remove nonpolars; the only question is how strong a solvent we need to elute our particular contaminant. If we are running a buffered mobile phase, we first must wash out the buffer. I usually keep a bottle of the same mobile phase minus the buffer on the shelf for washout at day's end before shutdown. Once we are in aqueous organic solvent, I switch to acetonitrile and wash the column for at least 6 column volumes (about 20 ml for a 25-cm analytical column). Watch the UV monitor for eluting peaks and a return to baseline. Reequilibrate with 70% methanol/water and run the four-standards mix. If it looks good, go back through the intermediate solvent to the buffered mobile phase, equilibrate, and return to your separation.

Be sure not to jump from buffer to pure organic or from organic to buffer. This can lead to buffer precipitation, plugging, and pressure problems. Always use a washout, intermediate solvent. Allow 6 column volumes for reequilibration; true equilibration takes as much as 24 hr, but this 6 volume equilibration is reproducible and sufficient.

If the late running peaks still run late or are spread, further washing is necessary. Directly after the four-standard mobile phase, I wash with 20% dimethyl sulfoxide in methanol. You may have to drop the flow rate initially to keep pressure below 4000 psi because of the mixture's high viscosity. The UV monitor will be of no use for monitoring peaks and a return to baseline since DMSO has very high absorption. After 6 column volumes, wash with standards solvent and reequilibrate, then run the four-standard mix.

The last resort is to wash all the way to hexane and back. Since aqueous solutions and hexane are immiscible, it is necessary to go through a bridging solvent(s). This means washing with one or more solvents miscible with both water and hexane. Common bridges are acetonitrile then chloroform, tetrahydrofuran (THF), or isopropanol. THF is probably the easiest and best bridge; its low viscosity allows rapid pumping. However, many people fear peroxide formation in THF, and decide to wash first with 6 column volumes of acetonitrile, then with chloroform, and finally with hexane; then reverse the process. Since each solvent must wash out the previous solvent completely, this is a very time-consuming wash. The isopropanol wash is also time consuming because of this solvent's high viscosity in water mixtures; it must be thorough since *i*-PrOH does not bridge as well as the other solvents. In any of these cases, you wash with each bridging solvent in turn (6 column volumes) until you reach hexane. You then reverse the process returning to the mobile phase for your column standards. The last step is to run the four-standard mix, then return and reequilibrate for the next sample.

It is better to pick a time convenient for you than to have to do this process on an emergency basis in the middle of a critical separation. I would have a tested column ready as a replacement. Replace the dirty column after washing out the buffer, cap it, and then wash the old column off line when you have more time. You never seem to wash everything off the column. After you have used a column for a while, you often will find a brown or black residue at the column head under the column frit upon opening even a freshly washed column. Do not worry about it if the column standards run correctly.

Washing with a bridging solvent seemed to correct about 80% of column contamination problems, but it could not cure the "disappearing peak" phenomena. In this case, the majority of the peaks remained unchanged, but a critical peak, usually in the middle of other peaks, will change retention time. Over a period of weeks or months it will merge with another peak, until it cannot be separated. Washing with solvent did not cure the problem. The change appeared to be an "α" change that pointed to a change in the chemistry of the system. After much work, it was tracked down to metal cation chelation. Speculation is that magnesium and calcium ions in the water (or from the glass in glass reservoirs) bind to free silanol sites on the packing, which changes its running character.

As we mentioned above, even endcapped packings have some free silanols, either left over from incomplete binding or by hydrolysis of the bonding. These give a reverse-phase separation a mixed mode nature. Most of the separation is due to the nonpolar partitioning bonded phase, but some of it comes from these ionizable, polar silanols. Metal cations form a pair bond couple and lock the silanol into the ionized form. The partition changes.

The answer suggests the treatment. Compounds that chelate metal ions should restore activity. EDTA proved unsuccessful because of steric factors, but oxalate at pH 4.0 succeeds in about 90% of the cases. The wash is made by adjusting the pH of 100 mM oxalic acid with 1 N sodium hydroxide. The column is washed with *6 column volumes* of oxalate, then with water until the effluent pH reaches the neutrality of your laboratory water. Do not overwash with this solution. Oxalate will attack the stainless steel tubing and extract iron if you wash longer. I know this from experience; a student in one of my classes did not listen when I told him 6 column volumes. He washed a column with oxalate overnight and had a reddish-brown waste container in the morning.

The last type on bound material is charged organic cations. They are usually of two sources: proteins and ion pairing reagents. They generally cannot be removed once they are on the column. The best treatment is either to prevent them from reaching the column or to dedicate a column to their use. If you must try and wash either type off the column, try using 70% acetonitrile containing 0.1% trifluoroacetic acid. Silanol has a pK_a around 1.8 and must be in the free acid form to release the cations. This solvent is used to solubilize peptides and small proteins, and might work. Realize, however, that you are walking a tightrope between removing the cation and the bonded phase.

Proteins are best removed from sample before injection, and various techniques will be described in the sample preparation section (Chapter 12) for doing so. If you must shoot crude sample-containing protein, use a guard column and change it often. A new guard column might be less expensive than the time needed to clean it, and, certainly, will be less expensive than a new column.

Ion pair reagents are used in separating charged compounds. They are charged molecules themselves and are used in fairly high concentration. Restoring a C_{18} column to initial conditions after using ion pairing reagents takes days of washing. These columns are usually dedicated to ion pairing runs. After use they are washed with solvent to ensure that the column's endfrit is free of solid, and the column is capped and stored until the next use.

6.4 Pressure Increases

The next column killer class is pressure increases. Most columns and packings can tolerate pressure of 12,000 psi and higher. Most new columns do not exceed 2500 psi when running the four-standard mix. If pressure rises to 4000 psi, you have a problem that should be dealt with. Be aware that gradient mixtures of solvents like methanol and water go through a pressure maxima that will approach 4000 psi at a 1.5 ml/min flow rate. I am talking about a change in system pressure that takes place gradually or all at once for no apparent reason and remains.

The first step is to locate the point of the pressure increase. Since most problems are column problems, we can simplify our task by "eating the elephant one bite at a time." Remove the column from the system and turn on the pump. If the pressure problem goes away, it was in the column. If not, it is in the system up to the column. I will deal here only with the column pressure problems; the system problems will be dealt with in Chapter 10 on troubleshooting.

There are three areas in a column where pressure increase can occur: the inlet frit, the outlet frit, and the column bed. The most likely source of problems is the inlet frit. It is the only filter between injected samples and the column bed, and is designed to collect anything bigger than 2 μm in size. When it does so, pressure increases. The more garbage it collects, the higher the pressure. You are left with the Fram oil filter alternative: "Pay me now or pay me later!" You can filter or spin particulates out of the sample, or be prepared to remove, replace, or clean the filter. Inlet filters can also be plugged by buffer precipitation caused by suddenly going from buffer to organic or vice versa. Step one is to replace the frit by opening the endcap, removing the old frit, and putting the new one on the column top. Did you remember to get replacement frits for your columns? Sometimes, usually late at night, you will find you do not have a new frit available. Instead, you have friends who just borrowed your last frit because they forgot to get them when they ordered columns.

In either case, if the frit plugs it can usually be fixed. Open the endcap and carefully remove the frit with the column in an upright position. (If you point it the wrong way you end up with white powder on your shoes.) Put the frit in a covered flask with 20% nitric acid (6 N) and sonicate it for 1–2 min. Carefully discard the acid, add distilled water, and resonicate. Keep washing with water until the water's pH reaches laboratory neutral. Replace the frit, blow the end-cap thread clean with a pipette to remove silica particles that can score column treads, and tighten the endcap.

If you reconnect the column, start the pump, and the pressure persists, you need to remove the outlet endfrit in the same way. (Remember the white packing on the shoes?) Outlet pressure is due to fines in the column collecting in this filter. Since they are silica, they can be cleaned by sonicating with 10% sodium hydroxide. Wash the base out repeatedly with water, replace the frit, and run the column.

If increased pressure still continues, then the column bed is probably plugged. If you can get any flow through the column, you may be able to wash out the problem. You must analyze the source of the problem. Bed plugs come from precipitated buffer and/or precipitated sample. Wash out the buffer with water and the sample with organic solvent. Sample precipitation usually happens only if you are doing preparative work with nearly saturated solutions. A column is a concentrator. It can supersaturate the solution at the head of the column, causing the sample to precipitate. Precipitated buffer occurs when you switch rapidly from buffer to an incompatible solvent. The crystalline buffer can end up in the frit or in the bed.

If flow is completely blocked, you will have to open the column, remove the frit, and bore out the plug with a flat-ended spatula. Be very careful: The plug is generally only 1–2 mm deep. The packing can be washed with solvent or water, drained, and pasted back into the column like you were healing an end void.

6.5 Column Channeling—Center Voids

The final type of column killer is the center void. When I first entered the field, this was always thought to be fatal, especially using the irregular 10-μm columns. The symptoms are that of the collapsing chromatogram. Everything could have been proceeding normally when, suddenly, you notice that the retention times are becoming less for your late-running peaks (Fig. 6.7). When you would repeat the separation, the problem would become worse. As it proceeded, even the early running peaks would become involved. Finally, everything is coming off in the void volume. If you called the company, they would tell you that the column was voided and would have to be replaced. If it was a brand new column that was shipped to you in this condition, they would probably offer to replace it. If not, then they might tell you that it would make great column packing for end voided column repair. They would also give you Standard Lecture #1 on protecting columns: "Do not jar the column, shock it through pressure changes or by jumping to immiscible solvents; do not reverse the column flow, or jump flow rates suddenly." All of these could cause voiding in perfectly good columns. In other words, tough luck. I have given this lecture many times myself, but no longer.

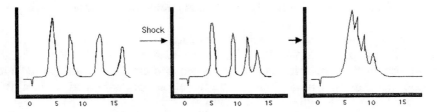

Figure 6.7. Effect of channeling—center void.

I am here to give hope to the masses. Voids can be healed! Well, not all of them, but many. The healing technique was discovered by accident, an example of serendipity. A novice chromatographer trying to get a job running an HPLC was told to run a standard mixture on a new C_{18} column and return to the interviewer with the separation. He proudly returned with a single peak, only to be told there were four in the mixture. The interviewer told him that the column was probably voided and needed to be returned to the manufacturer, and warned the interviewee not to do anything to invalidate the warrantee. While trying to get some practice with the voided column, the novice dropped it on the floor; in a panic he hooked it up backward. Next, he accidentally turned the flow to 10 ml/min instead of 1 ml/min. After he got the flow under control, he shot a four-standard mixture only to find that it came off as four peaks instead of the expected single peak of a voided column. He had healed the center void. The interviewer recognized what had happened and passed the word through the company. The trick was tried on other voided columns and proved successful in 13 of 14 tried. When I first heard of the technique used to heal columns I was skeptical. I went to my demonstration system, ran my standards, and got four peaks. I intentionally voided the column by running at high pressure, then suddenly dropping the pressure. The third time I did this I got a single peak center voided chromatogram. Following the protocol, I removed the column and banged it against the counter.

The method practiced by a major industrial account was to disconnect the center voided column, grasp it in one hand and rap the counter with it twice, reverse the column, and do the same with the other end (obviously, not hard enough to bend the column!). Next, they hooked it up backward and ran it at a high flow rate for a minute or two. They then ran the four-standard mixture. From then on the column was run reversed. It is possible that an end void may be formed and must be repacked, but the column bed should be restored. When I repeated this technique and reinjected my standards, I got four well-resolved peaks.

A center void probably occurs because small wall and bed voids link up under pressure changes and shock to form a channel. The channel is a path of least resistance and diverts the flow from the bed. This effectively removes this part of the column from the separation and gives a shorter column and shorter retention times. The whole column is eventually channeled and you have a center void the length of the column. If you have ever run a gravity-fed, packed glass column, you have probably seen this phenomenon if you accidentally let the column run dry.

This crazy-sounding repair technique probably works because a column is more densely packed at the outlet than at the inlet end. This represents a packing reservoir that can be loosened and washed into the center void. I would not recommend using it on anything but a hopelessly center-voided column. By the same token, I would not recommend reversing the flow of a column needlessly. Occasionally, reversal has caused columns to void. If you have already used up the "packing reservoir" by reversing the column it may not be available to fix a void if it does occur. This is what happened in the unrepairable fourteenth column mentioned earlier. It had already been reversed before the void opened and the repair technique failed on the second reversal. Why waste a resource?

6.6 Normal Phase, Ion Exchange, and Size Columns

Most of the mentioned troubleshooting tools will work with other silica-based columns. With normal-phase columns, you obviously need not worry about bonded-phase removal, but silica still dissolves at high pH and high salt. Polar materials like some proteins adhere very tightly and require high acid and low pH to be washed off. My customers were able to run normal-phase separation of phospholipids using MeOH, acetonitrile, and concentrated phosphoric acid, conditions that would be totally unacceptable on a bonded-phase column. Particulates can result in pressure problems and are removed from the frit in the same manner as with bonded-phase column. Center and end voids can be repaired in the same fashion.

Ion-exchange columns on silica show exactly the same problems as other bonded-phase columns: bonded-phase loss, column packing loss, bound organics, pressure problems, and end and center voids. In addition, they exhibit binding problems specific to their function. Strong ion exchangers can bind almost irreversibly to strong ions of the opposite charge (i.e., quarternary amine columns with sulfonic acids such as taurine). To break this electrostatic attraction it is necessary to either neutralize one or the other of the ions or displace the bound ion with very high concentrations of a counterion, such as salt. Neither the quarternary amine nor the sulfonic acid can be neutralized in our example without wrecking the bonded phase. Salt (1–2 M) dissolves the packing material when used as a wash over a period of time, but it is the only solution to this problem. A better solution is not to use strong ion exchangers; weak ion exchangers, for instance, a DEAE column that uses secondary and tertiary amines, can be cleaned by using high pH control and a saturation column. Remember that when you are using amine columns you must use deoxygenated solvents or face amine oxidation. In recent years, a number of ion exchangers on very rigid polymeric supports have emerged. They can tolerate reasonably high pressures and pHs from 2 to 13, and even briefly 1 to 14. They are ideal solutions to the problems seen with silica-based columns and should displace silica in ion exchange in the near future.

Size-separation columns on silica show all of the bonded-phase problems and can be treated in much the same fashion. They show problems specifically related to their operation. Pore size is critical to their operation. Anything that blocks pores changes their operation. Adhering materials, such as nonpolar contaminants, proteins, and detergents, can have a major effect on exclusion/inclusion ratios. Pressure fragility is very common, especially on the larger pore size columns used for large exclusion/inclusion separations. The TSK4000sw column appears to be packed with very fragile Christmas tree ornaments and should be handled accordingly. We had a major outbreak of crushed TSK3000sw columns a few years ago. On one, a customer called me and said that the first time he ran the column the pressure went up to 4000 psi and stopped the pump. I gave him a new column, took back his old column, pulled the endcap, and tried to feel the column head with a glass rod. There was almost nothing in the column; I almost lost the glass rod. The column had crushed and packed into the outlet end. TSK3000sw columns are usually not that fragile! I inquired around the country and found 12 from the same lot that had had the same problem. The manufacturer finally admitted they had been shipped across the Northern Pacific in the dead of winter in 10% MeOH. The mobile phase had frozen, expanded, and crushed the column packing. They started shipping in 50% MeOH and, to my knowledge, the problem never happened again.

CHAPTER

7

Modifications of Partition Chromatography

7.1 Reverse Phase

So far we have looked at reverse-phase separation using simple solvent mixtures. When we carry out separation of charged compounds or ones that can be ionized we often run into two problems: tailing or poor retention.

Chromatographic peaks are asymmetric and tend to broaden or tail off on the side away from the injection point. As a result, peaks tend to contaminate longer retaining neighbors. Extreme tailing, which is always due to some type of poorly resolved equilibration, must be dealt with before separations can be achieved. One of the most common causes of tailing is partial ionizations, either of the column surface or the sample in the mobile phase. For instance, at the pK_a of an acid, the carboxylate form and the free acid form are present in equal amounts. If you run a column buffer at the pK_a and try to separate this acid from other compounds, the result will be a badly tailing peak as the column tries to separate the two equilibration forms.

Another problem is compounds that are too soluble to retain on the column, and elute unresolved in the void volume. Modifications of either the sample ionization or of the surface nature can increase retention and, therefore, resolution. In this chapter we will study modifications of the column or mobile phase that will allow us to improve our separations.

7.1.1 Ionization Suppression

Buffers are used in HPLC to control the ionization of one or more molecules in the solution so that they will separate as sharp bands. The key to understanding ionization is to understand pH and pK_a.

The pH of a solution is simply a measure of the hydrogen ion concentration and represents the degree of harshness on either the acid or basic side. A pH of 7 is neutral and represents the mildest condition. As we go toward a lower pH the hydrogen ion concentration increases and the solution becomes more harshly acidic. Starting at 7 and going toward a higher pH the hydrogen ion concentration decreases and the solution becomes a harsher base.

pK_a for each ionizable function on a molecule is the pH at which equal concentrations of the ionized and free form exist. Organic acids have pK_a around pH 4.5 and amines have pK_a between pH 9.0 and 10.5. Below 2.5, organic acids exist mainly in the protonated, free acid form. Above 6.5, the proton is removed and, mostly, the carboxylate form is present. As a rule, try and stay 2 pH units above or below the pK_a of the compound being separated in your HPLC column. The worst tailing seems to occur directly at the pK_a. Also be aware that some compounds have more than one ionizable functional group and show more than one pK_a.

Since buffers control pH best at their pK_a, pick one close to your desired pH. The most common buffer used in HPLC is phosphate. It has two usable pK_as, 2.1 and 7.1, and is UV transparent. A 100 mM solution of phosphate precipitates in solution of > 50% MeOH or 70% acetonitrile. Other buffers in common use are acetate, pK_a 4.8, formate, pK_a 3.8, and chloroacetate, pK_a 2.9; all absorb in the UV below 225 nm. Sulfonate, pK_a 1.8 and 6.9, could be substituted for phosphate when analyzing a mixture of organic phosphates where it may be desirable to cleave the compound and do an inorganic phosphate analysis.

The other factor to consider is the effect of ionization on solubility. Ionized forms are more soluble in aqueous solvents. If you need to increase a compound's retention on a reverse-phase column, force it into its nonionized form. For small organic acids, it is best to run your separation at pH 2.5 with phosphate buffers or use 100 mM acetic acid, which give you a pH of 2.9. For large acids containing massive nonpolar substituents, it might be better to operate at pH 6.5 and take advantage of the decreased retention time for faster chromatography.

Amines pose an interesting problem for ionization control because their pK_a are so high that they are usually ionized at any pH tolerated by the silica column bed. This makes them very soluble and hard to resolve on a reverse-phase column. It is possible to force them into the free amine form by using mobile phases at pH 12, but be sure to use a saturation column and change it often.

7.1.2 Ion Pairing

Amines have traditionally been separated by using ion pairing reagents. These are countercharged organic molecules, such as hexane sulfonate, that are added in excess (typically 30–100 mM) to the mobile phase. One theory says that they form an "ion pair" with the amine in solution that becomes one long nonpolar pseudomolecule with a masked charge couple in the center. The pseudomolecule then partitions with the bonded phase as if the charges did not exist.

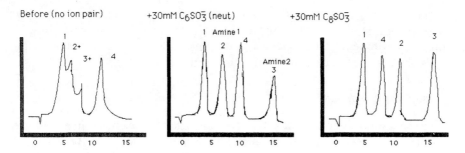

Figure 7.1. Using ion pairing reagents.

Instead of eluting at the void volume like the ionized amine, the psuedomolecule is retained longer than even the free amine. An alternate theory of ion pair action says that the ion pair reagent first interacts with the bonded phase forming a nonbonded ion-exchange column. This modified bonded-phase column then interacts with the compounds in solution through a mixed partition/ion-exchange mode.

The longer the nonpolar chain of the ion pairing reagent, the longer this retention (Fig. 7.1). This allows us to position the retention time of an amine in a separation by controlling chain length of the ion pair. A very interesting observation is that a 1:1 mixture of hexane sulfonate and octane sulfonate gives a single amine peak retaining half way between the peaks formed when either sulfonate is used alone. This 1:1 mixture has the same retention as the amine reacted with heptane sulfonate. Neither of the two proposed theories of ion-pairing interactions explains why this mixed ion-pairing reagent does not form at pair of peaks or a badly tailing peak for each compound formed.

Ion-pairing reactions can also be carried out using quaternary amines as countercharges for organic acid and organic phosphate samples. Generally, pH control is the preferred technique for acids, but ion pairs occasionally give better chromatographic positioning or solubility control.

If ion-pairing reagents are used in gradient runs, they must be added to each solvent in equal amounts to prevent baseline drift during the run. The pairing reagent should be transparent at the wavelength being used, if possible.

7.1.3 Organic Modifiers

The other ionization that causes tailing in reverse-phase separations is ionization of the packing surface. As previously mentioned, there is a small percentage of free silanols. The older the column, the more likely that more of these free silanols will be present due to packing material hydrolytic degradation. These are available to react with amines in the mobile phase through an ion-exchange interaction. This effect can be greatly overcome by adding 5 mM nonyl amine to the mobile phase during equilibration and during chromatographic runs. The amine function of this competing base or organic modifier ties up the free silanol presenting a nonpolar surface to sample amine in solution. The competing base effect is very dramatic at low pH, but also shows some peak sharping when used at pH 10 with a saturation column.

7.1.4 Chelation

The separation of compounds that serve as ligands for chelating metals can often be enhanced by adding the metal salt to the mobile phase. If the ligand 1 coelutes from the column with a nonligand, adding a soluble chelating metal cation will increase the ligand's solubility and decrease its retention time, pulling the two compounds apart (Fig. 7.2). An immobilized chelating metal can be created by first forming a complex of it with a nonpolar molecule possessing a ligating functional group, then saturating the reverse-phase column with the complex. This nonbonded complex can be used to make a separation by tying up the compound in the mixture, which acts as a ligand causing it to run slower

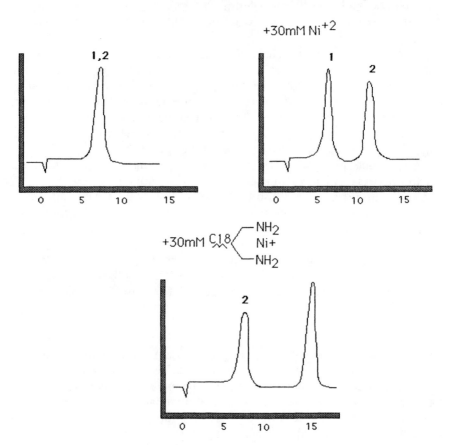

Figure 7.2. Using chelating agents.

than the nonligand component, again pulling them apart, but with a reversed order of elution.

This has recently been taken one step further and commercial bonded-phase chelation columns have been offered for sale. The interest in these separations has risen because the chelation with metals such as Ni and Zn is asymmetric and allows the selective separation of optical isomers present in amino acids, peptides, proteins, and carbohydrates. One optically active form of a class of compounds usually sticks tighter than the other. When you prepare Gly diastereomers of amino acids and run them on an immobilized NI column the L-amino acid diastereomer of each pair comes off faster than the D-amino acid form.

7.2 Acidic-Phase Silica

I have discussed normal-phase separations on silica and hydrated silica columns in which polar compounds retain and nonpolars elute. Separations have begun to appear in the literature using an acid-phase silica column for the separation of phospholipids.

These represent a new type of normal-phase partition separation. As much as 1–2% sulfuric or phosphoric acid is added to the mobile phase, forming a protonated hydration shell around silica. The phospholipids are soluble in nonpolar solvent, but differ from each other in polar functionalities, such as sugars, alcohols, amino acids, or varying numbers of phosphate groups. By going to very harshly acidic conditions, even phosphate groups can be forced into their protonated form, allowing them to be separated off the silica surface by nonpolar mobile phases. With no bonded phase on the silica to be affected by acid hydrolysis, pH can be kept very low. Everything can be stripped off the column with acid water, then washed back up to nonpolar solvents and the column requilibrated for the next injection. The only reaction of this type that I have seen used a mixture of An and MeOH in phosphoric or sulfuric acid. Obviously developed from a TLC separation, since it contains both MeOH and acetonitrile as well as strong acid, it could not be run as a gradient at lower wavelength needed to detect phospholipids because of the absorbing MeOH. It would be interesting to investigate some acetonitrile to water gradients in phosphoric acid. These would be transparent to 195 nm and good for resolving very polar components. If the compounds are being saved and cannot tolerate long exposure to concentrated acids, they should be run into buffered solution as soon as they elute.

7.3 Partition Mode Selection

How do you decide when to choose a reverse-phase instead of a normal-phase column or an intermediate-phase column such as a cyano column? Reverse-phase columns are chosen about 70% of the time, so most compounds can be separated by this partition mode. What in the make up of the compound being separated selects one column over the other?

We have already mentioned solvent solubility. If the compounds are not soluble in nonpolar solvents, there is little chance we will be able to separate them on a normal-phase column. The operating solvent ranges, however, are fairly wide on both columns, as we have seen, and a solvent can usually be found that will dissolve our compound and allow it to be run on the column.

Column selection often has to do with which area of a molecule contains the differentiating portion. Two compounds to be separated may vary by substitution on a benzene ring. Another pair may vary only by a polar functional group. Again, like attracts like. The working rule is to select a column in which the variable parts of two molecules point toward the column.

In the first case, the variation was in a nonpolar side chain, say, for instance, *ortho-* or *para*-toluic acid. We want the nonpolar end pointing toward the column, so we select a C_{18} reverse-phase column. We buffer the mobile phase to 7.5 to help orient the polar benzoic acid group toward the mobile phase and the substitutions on the ring toward the column surface.

In the second case, we have the same nonpolar side chain, but differing polar functions, say *p*-methylphenol and *p*-toluidine. We want the phenolic and anilinic functions toward the column, and, therefore, you would select a normal-phase column. The nonpolar solvent attracts the aromatic methyl substituents.

One other empirical rule. For some reason, positional isomers seen to be best resolved on anhydrous silica columns. I cannot offer a good reason why this is so. Separation of *cis-/trans-* and axial/equitorial isomers seems to proceed best on these normal-phase columns.

7.4 Hydrophilic Separations

Hydrophilic size separation columns for use with aqueous samples have recenty become very popular in purifying proteins and carbohydrates. Protein separation columns are available on both silica and polymeric supports. It is surprising that the best of these protein purification columns in terms of resolution and in recovery of native protein are silica-based columns. One would expect that protein release from silica would be a real problem. It certainly is in many other silica columns. These columns, however, especially the TSK family of columns, give excellent recovery. I have talked to other column manufacturers who have investigated this problem. They say that when you remove the bonded phases from these columns they appear to be identical to bonded phases from a number of other columns for protein purification. These bonded phases are primarily diol ether polymers, very hydrophilic, but of intermediate polarity. Some modification of the underlying silica appears to give the TSK columns their unique ability to release nearly all protein placed on them.

Like the size-separation columns discussed earlier, they are graded by the molecular weight ranges of their exclusion and inclusion volumes, but this is very deceptive. Globular proteins and many enzymes tend to fold in and wrap around themselves causing them to run smaller than you would expect from their molecular weights. If they are straightened out with denaturing mobile phases containing sodium dilauryl sulfate (SDS), they show a fairly linear relationship to molecular weight. A column with a molecular weight range of 10,000/40,000 will drop to 8000/25,000 in SDS. Columns exist with exclusion of 2.5 million; work continues to extend the range high enough to separate intact nucleic acids, but these very large pore packings are very fragile. The lower limit at the moment is around 8000 with silica, but efforts are being made to extend sized-separation ranges down to include large peptides (1000 MW). A customer of mine claims to be able to separate decapeptides using the smallest of the polymeric, hydrophilic size columns, the TSK-2000 pw. This column is sold as a carbohydrate size-separation column and he grabbed it by mistake when trying to separate these peptides from a protein mixture.

I have mentioned that solvent affects the separation of proteins. For recovery of native proteins and enzymes, mobile phases are selected to stabilize structure and preserve activity. Mobile phases very much resemble enzyme assay conditions. I routinely make separations in 100 mM Tris-phosphate buffer at pH 7.2. Metal ions and sulfhydryl stabilizers such as dithioglycerol are often added. Chromatography is sharpened with salts, phosphate, sulfates, and, best of all, 150 mM sodium chloride. The latter is unfortunate because it erodes the packing and corrodes the exposed metal surfaces. We can protect the system hardware as we will discuss in the next chapter, but there is little we can do about the column wall except go to glass or plastic. The packing has to fend for itself and end voids are common in these columns. Even worse, it is very difficult to buy any quantity of packing material for topping up columns. Of course, you can always buy a new column and sacrifice the old one to provide topping material.

One mobile phase additive—glycerol—serves a double purpose. Up to 10% glycerol is often added to stabilize protein activity. It also serves to decrease interaction of glycoproteins with the diol column packing and make the glycoproteins come off faster. A glycoprotein can be made to run slowly in a glycerol-free mobile phase, then reinjected into the same column containing 5% glycerol, which makes it elute more quickly than compounds with which it formerly coeluted. This can produce a two-step sequential analysis on a single column. Isopropanol is also said to have a similar effect.

Crude plasma quickly fouls the protein separation column with lipids, usually after only three injections. These lipids can be removed by washing the column with water, then with 20% DMSO/MeOH, and finally with water again before returning to buffer.

Enzyme purification is not the only job of the size-separation column. These columns also serve as protein preparative columns and will accept as much as 100 mg of protein at a shot. When this protein is to be used for structure determination, detergents can be used in the mobile phase to increase large protein solubility. Nonionic detergents will often give enzyme activity back on dialysis, but the ionic detergents seem to finish activity off by making permanent structural changes. The ionic detergents act very much like ion-pairing reagents in partition work and are very difficult to remove. It is generally better to dedicate a column rather than to take a chance on losing your next enzyme preparation to a dirty column.

Another series of hydrophilic size-separation columns are based on cross-linked polymers. They are sold as carbohydrate size-separation columns and will separate a polymer series from each other, but will not separate the monomeric isomers (i.e., glucose from galactose). These columns also work for proteins and peptides. They have the same diol type of bonded phases as the silica-based columns, but do not show as broad a molecular weight range or as high a resolution. Because they are polymer based, they will not take pressures over 1500 psi and should not be cleaned with organic solvents. They show considerable promise for separation of heparin and chondroitin sulfate-type polysaccharides. Again, detergents and glycerine can be used to increase solubilities and to control sample interaction with the bonded phase. Heated mobile phase speeds equilibration and improves peak shape and resolution.

8

"Nonpartition" Chromatography

So far we have dealt primarily with partition separations, which represent about 80% of HPLC runs. Now we turn to other separation modes. Size separation makes up another 15% and ion exchange the remaining 5%. Silica and polymer column supports are available for both separation modes. Except for the carbohydrate separation column, almost all HPLC-based ion exchange is carried out with silica-based columns.

Size separation uses both silica- and polymer-based columns. Even though both of these techniques are supposed to be free of partition effects, in the real world, these are bonded-phase columns. To use them successfully, you must not only understand the basic separation mode, but be able to correct, eliminate, or take advantage of the partition effect.

8.1 Ion Exchange

Ion exchange relies on charged columns attracting oppositely charged molecules from the mobile phase, then releasing them in inverse order of their attracting strength. Two types of ion exchangers exist: cationic and anionic, named for the types of molecules they attract. Each can be divided into two subtypes: strong and weak, depending on the type of ionization of their column bound functional group. Strong groups are ionized at all working pHs, while weak groups can be either charged or uncharged depending on the pH. Three techniques can be used to remove attracted counterions:

1. competitive displacement by a mobile phase salt ion,
2. pH control of the attracted ion's charge, and
3. pH control of the column's charge.

8.1.1 Cationic: Weak and Strong

Strong cationic ion-exchange columns have an aromatic sulfonate connected through an aliphatic side chain to the silica surface. In the pair-bonding stage, cations injected into the mobile phase are attracted to the column, while the column counterion, neutral compound, and anions are eluted in the void volume. In the second step, salt cations from the mobile phase attack and competitively displace the sample cation from the pair bond. The stronger the bond, the tighter the hold of the sample ion on the column and the longer the retention time. These two stages obviously run simultaneously and repeatedly as the compounds move down the column and are eluted as separate bands. This salt displacement can be supplemented with salt gradient elution. Increased salt pushes bands off faster and somewhat sharpens them. Many ion-exchange chromatographers prefer to run isocratic mobile phase, shoot the sample, then immediately shoot two injectors full of a very concentrated salt solution. They claim they obtain sharper bands and better separation.

We are not limited to using only salt displacement to remove the sample cation. If it is a weak cation, the charge can be removed by running a pH gradient to high pH. Separation will be in order of pK_b with the lowest pK_b coming off first. Since many of the naturally occurring cations are amines, it will be necessary to go to pH around 12 and a saturation column will be required. Because of buffering effects both of added buffers and column surfaces and even injected compounds, pH changes are never as precise as organic solvent gradients used in partition separations.

Strong cationic exchangers attract all cations in the mobile phase. If it is a strong cation (i.e., a quaternary amine), and the compound it attracts is a strong anion, they will form a very tight pair bond that can be broken only by long washing with high concentrations of salt, which erodes the packing. This can be avoided by using a weak cationic ion exchanger that has an alkyl carboxylate bound to the silica. The weak exchanger forms a weaker pair bond with the strong cation allowing salt elution. If necessary this column can be stripped of cations by going to pH 2.0. The weak ion exchanger is now in the acid form and must be regenerated by reequilibrating at pH 6.5. The carboxylic acid column is the functional equivalent of the carboxymethylcellulose (CM) column used for protein separation in open column work.

One other cationic column in common use is the carbohydrate column. This column has a sulfonate bound to a polymeric support and is used pair bonded to a large cation, calcium or lead. It probably separates carbohydrates by a partition rather than an ion-exchange mode. The column is run in water at 80°C to speed mass transfer and decrease viscosity. It is very fragile and should not be run over 1500 psi. Even accidentally jumping the flow rate can rupture the column. Small amounts of organic alcohols sharpen peaks, but do not let the organic content exceed 20%. Polysaccharides come off the column first, followed by trisaccharides, then disaccharides, and finally simple sugars. This column will resolve among monosaccharides and separate sugar alcohols from each other.

8.1.2 Anionic: Weak and Strong

In strong anionic ion exchangers, the quaternary amine is attached to the silica surface by an alkyl chain, and the attracted anion comes from the mobile phase. In the pair-bond stage, the anion attaches while neutrals, cations, and the column's counterion elute in the void volume. Salt displacement releases anions with the weakest ion coming off first. Strong anions poison the exchanger, but can be washed off with very strong salt. Weak anions can be removed with salt displacement, but also by using pH gradients from pH 6.5 to 2.0. The free acids are eluted in order of their pK_as with high pK_a off first.

Weak anionic exchangers use bonded phases with either primary, secondary, or tertiary amines as the function exchanger. It forms weaker bonds with strong anions and can be cleaned by going to high pH using a saturation column to form the free amine form of the anion exchanger.

Weak amine columns, whether primary, secondary, or tertiary, will all oxidize in solvents containing oxygen and need to be protected by nitrogen-purged vacuum-treated solvents as mentioned in Chapter 6. It may seem inconvenient, but column life will be 3 months or less without it.

Both cationic and anionic silica columns occasionally need to be repaired. If you have the same packing material, make a paste of it with mobile phase and paste it in. If the same packing is unavailable, use cyanopropyl packing for small repairs. If necessary these columns can be washed with water, then with 20% DMSO/MeOH, and with water; finally, reequilibrate with buffer.

Do not try to open or repack polymeric columns. They are usually under some pressure and come out of the tube like toothpaste. The column is then of no use. Polymeric columns are usually packed in one solvent, then switched to a second solvent, which causes the packing to swell and squeeze out voids. They are then designed to be run in the second solvent. Polymeric ion exchangers are usually run at elevated temperature. This serves two purposes: It decreases mobile phase viscosity, thereby reducing operating pressures, and it speeds equilibration of the samples with the ion exchanger, which is usually very slow with these columns.

8.2 Size Exclusion

Although the oldest type of columns, these are presently the most popular columns because they can separate large biological molecules such as proteins and nucleic acids. By means of controlled pore sizes, they separate compounds by means of their molecular size and shape. The resolution achieved by a size column is not nearly as great as that shown by ion exchange or by partition. You will generally need a 100% difference in molecular weights to achieve a separation; a size column can separate monosaccharides from disacchardies. Partition can separate on the basis of a proton up or down out of 13 protons on a compound.

Although we describe these as molecular weight columns, the separating parameter actual is their Stokes radius. The shape and folding of the molecule under differing solvent conditions affect their maximum radius and, therefore, their retention times. Only when extreme denaturing conditions are used to unfold and force all molecules into the same shape are we able to obtain a direct molecular weight relationship.

Both polymeric and silica-based columns are in common use. The polymeric columns are heavily used in the analysis of synthetic polymers and plastics where organic solvents are required. Silica-based columns with hydrophilic bonded phases are used to separate aqueous solutions of macromolecules. Finally, polymeric size-separation columns with hydrophilic phases are available for separation of both polysaccharides and proteins.

8.2.1 Organic Soluble Samples

Polymer samples are size separated by dissolving them in an organic solvent, such as THF, then passing them through a linked series of sizing columns. These crosslinked polymer columns vary in pore size with each column graded by its inclusion/exclusion ratio. Ratios start at less than 100 and go to well over 20 million Da; a typical range for a single column would be from 30,000 to 200,000 Da. The separating range of each column is determined using polystyrene standards of known molecular weights. While each column may have a narrow separating range, connecting an ascending series together with the smallest next to the injector produces a combination with a range from the smallest packing's inclusion to the largest packing's exclusion. The range will be continuous if each column's exclusion limit overlaps its neighbor's inclusion.

Switching solvents is very poor practice because of bed swelling and shrinking from solvent to solvent. A solvent is usually selected with a broad solubility range; the system is turned on and allowed to equilibrate for 24 hr and then kept in a constant flow recycle mode until needed. When a sample is to be shot, flow is switched out of recycle, the chromatogram is run, and the system is then immediately returned to recycle. The pump is left on at all times.

Since polymers are not discrete compounds, but are instead a range of compounds, the chromatograms produced are not a series of peaks, but a continuum with peaks. Measurements are generally made with a refractive index detector and the amount of material present at various points in the trace is measured. Early running components are high molecular weight and give information about stretch and flexibility. Later runners are smaller and give information about leaching, solubility, and brittleness. The chromatographer is often less interested in determining molecular weight distribution, then in getting a "fingerprint" of the particular polymer. This distribution fingerprint can provide information on unreacted monomer and degree of polymerization, and serve as a batch-to-batch quality control device.

Very high-density polymers are run in a special high-temperature HPLC. This device can automatically dissolve, filter, inject, and run these samples at 200°C in a solvent such as chlorobenezene.

8.2.2 Aqueous Soluble Samples

Hydrophilic size-separation columns for use with aqueous samples have recently become very popular for use with proteins and carbohydrates. As mentioned in the last chapter, protein separation columns are available on silica and polymeric supports.

Size columns tend to dilute the sample shot into them, unlike partition, ion exchange, or affinity columns, all of which tend to concentrate samples placed on them. To obtain maximum effect, size columns need to be tall and thin to allow maximum time for compounds to interact with column pores without diffusion upsetting the separation. Since they are generally the poorest column type for achieving resolution they have two main uses: First, they are used as a first column to tear compounds in to size groups before going to a concentrating separation mode; second, they are the final column of choice to remove buffers and salt from elution fractions from other separating modes. Desalting columns based on Sephadex G-25 are a much faster and more complete method of removing salts then either dialysis or molecular weight filtration membranes.

8.3 Affinity Chromatography

Affinity chromatography is a technique more commonly associated with low-pressure columns and packings. The column must be prepared for each new separation.

8.3.1 Column Packing Modification

First, the target compound to be bound to the column must be identified, isolated, and activated. In some cases, the column packing is purchased already activated. Once the target is chemically bound to the column the packing must be slurry packed into the column. Fortunately, these columns concentrate and bind the substrate so they can be broad and narrow and easily packed.

In certain cases, affinity columns can be used to fractionate within a class of bound materials, for instance, antibody columns have been used to separate the various subtypes of IgG. In this case the packings are microporous, heavily crosslinked polymers and benefit from HPLC operating conditions. Eluting conditions are usually step gradients of buffers with different pHs. The last step of a protein G column is at a very acidic pH and the sample is eluted into buffer that quickly raises the pH to prevent protein denaturation.

8.3.2 Enzyme "Substrates" and Dye Substrates

Proteins and antibodies are natural substrates for affinity columns because of the nature of the enzyme recognition site and the antibody–antigen interaction sites. They have a three-dimensional shape and electrical charge distributions that interact with only specific molecules or types of molecules. Once these substrate sites are identified, molecules can be isolated or synthesized with the key characteristics and used to build affinity supports. These substrates are often bound to a 6-carbon spacer so that they protrude farther away from the packing surface toward the mobile phase and are therefore more available. Certain natural and synthetic dyes have been found to serve as substrate mimics for a class of enzymes called hydrogenases and have been used to build affinity columns for their purification.

8.3.3 Chelation and Optically Active Columns

Vitamins are known to act as cofactors for certain enzymes and have been studied as affinity targets to be used in purification of those enzymes. Another class of general cofactors for enzymes is the metal ions that many enzymes require for their activity. They lead to a generalized affinity packing, one with a bound group that can chelate a number of transition metal ions. Typical of these is a long-chain organic molecule with branching chains, each of which ends in an amino group. These "amino finger groups" can serve as ligands for the metal ion. A tridentated (three finger) molecule can cup the metal and allow ready access from only one side. This type of target group has been used to resolve racemic mixtures of optical active compounds on commercially available columns. One optical isomer of an isomeric pair can approach and bind the complex better than the other leading to a separation. Elution can be made using salt or pH to break the complex. For very tightly bound material, the natural substrate will generally exhibit a tighter fit and serve as a displacer. For the metal ion affinity column, the binding can be broken with strong metal chelators such as EDTA.

9

Hardware Specifics

HPLC separation occurs when a mixture of compounds dissolved in a solvent may either stay in the solvent or go onto the packing material in the column. The choice is not a simple one since compounds have an affinity for both the solvent and the packing. In fact, separation occurs because different compounds have different partition rates. Left to themselves each compound would reach its own equilibrium composition in the solvent. However, we upset conditions by pumping fresh solvent down the column. The result is that the components with the highest affinity for the column packing stick the longest and wash out last. This differential washout or elution of compounds is the basis for HPLC separation, a very simple technique that often requires complex equipment for its execution.

9.1 System Protection

The simplest HPLC system is made up of a high-pressure solvent pump, an injector, a column, a detector, and a data recorder (Fig. 9.1). The high pressures referred to are of the order of 2000–6000 psi. Since we are working with liquids instead of gases, high pressures do not pose an explosion hazard. Leaks occur with too much pressure. The worst problems to be expected are drips, streams, and puddles.

An isocratic system is used with single solvents, a premixed solvent mixture, or step gradients. It has the advantage of needing a single pump, no mixer, and no controller. Because of this, isocratic systems are also simpler and less expen-

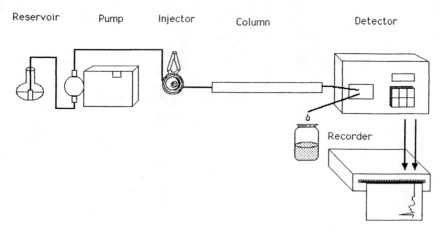

Reservoir   Pump   Injector   Column   Detector

Recorder

Figure 9.1.   Isocratic system components.

sive then gradient systems. With an isocratic run there is no necessity to reequilibrate to initial conditions before making your next injection. Many HPLC runs do not require a gradient. Isocratic systems run faster because they do not require reequilibration. They are less expensive, since they do not need all the fancy gradient hardware. At present, an informal survey of users has shown me that gradients are used approximately 20% of the time, although gradient systems are purchased at a much higher rate.

9.1.1   Filters, Guard Columns, and Saturation Columns

The tubing used to tie everything together is of extremely small diameter and is easily plugged. Packing beads are made of very fine particles and must be protected with a series of filters and guard columns. The usual first filter is a 10- to 30-μm pore size sinker on the end of the solvent line in the solvent reservoir. This filter is probably redundant since HPLC solvents should be filtered through at least a 0.54-μm filter before use. Some pumps may also have a 5- to 10-μm filter before or in the pumping/check valve system. Too many filters before the pumping chamber can lead to problems because they provide a resistance to flow and may cause air bubbles to form in the pump head. If the solvent sinker gets plugged with solids or rusts it can restrict solvent flow and lead to pump starving. This can be determined by using a graduated cylinder to measure the pump output versus the selected pump flow. I also sold a high-pressure mixing system that used a magnetically stirred high-pressure mixing chamber than was equipped with a filter. We were not aware that the filter was present until we developed a pressure problem and traced it back to the mixer using reverse diagnosis techniques.

We have already discussed the use of a saturation column between the pump and injector to protect the main column against very high pH buffers. Another function of this column is to provide a filter (the column inlet and outlet frits) before the injector. I have seen unusual pumps that flaked Teflon® off the pump seal; this ended up plugging the injector. An in-line filter would have prevented the problem.

The only filter between the injected sample and the column bed is the inlet frit of the column (pre-column). Because of this it is very important to filter *all* injected samples. I had a customer who got a pressure increase every time she injected her standards. It never occurred to her that she might need to filter standards. When we dissolved her standards in solvent in a test tube and swirled them, you could see an opalescence. Centrifugation brought down a white pellet. The standard had evidently been decolorized with charcoal and filtered through a Celite bed. Celite fines probably passed through the filter paper with the solution and were trapped in the compound on final evaporation.

A guard column placed between the injector and the main column is often used to trap contaminants that might stick on the main column. The guard column inlet filter now becomes the first filter after injection. The advantage of a guard column is that it is short and can be reversed without voiding when washed into a beaker with solvent flow from the HPLC pump. Particulates trapped in the guard column inlet frit are also washed out at the same time. Once washed the guard column can be inverted and placed back into service.

9.1.2 Inert Surfaces and Connections

The stainless steel used in HPLC lines, fittings, and other wetted surfaces is corroded by exposure to halide salts. In systems that require routine operation with salt, protection is available in two forms. If less than 200 mM salt are to be used, the pre-column and column can be removed, a column bridge can be added, and the system can be treated with 20% nitric acid followed by water washing. This *pacification* treatment is also used to remove precipitated buffer crystals and organic precipitation in lines and injector loops for up to 1 month. If a more concentrated halide operation is foreseen, it might be worthwhile to purchase a system with inert wetted surfaces. These consist of titanium pump heads and flow cells and *PEEK*-type pressure-resistant tubing and fittings. Titanium construction adds surprisingly little to the cost of a system, but does present some problems. Titanium is very hard and brittle. Titanium lines are available, but should be avoided. They are hard to cut without breaking, and compression fittings do not bind well and can split lines while being attached. Two problems I have seen with inert systems should be avoided: First, lines going into inert flow cells are soldered in place with nickel or silver solder. These metals leach out with corrosive solutions and appear in the effluent. The second problem is with pump pressure sensors (transducers). These flow-through devices are inside the pump and are often ignored or forgotton by inert pump manufacturers. If inert transducers are not used during inert pump construction, they will corrode and can eventually collapse or begin leaking inside the pump. I have seen this happen and it is not a pretty sight to see the whole inside of a $4000 pump corroded with strong salt solution.

9.2 Pumping

Pumps are basically devices for pulling in solvent, pressurizing it, and driving it out through the injector, column, and detector. As described earlier, it does this with a plunger driven through a seal into a pumping chamber (Fig. 9.2).

Inlet and outlet check valves ensure one-way solvent flow. Problems arise when the plunger must be pulled back to ready itself for the next stroke. The pressure drops until the plunger starts forward again. This results in pulsation, which causes variations in solvent delivery flow, and, more importantly, variations in pressure to the column. The column acts as a pulse dampener, which can easily be seen by watching the baseline of a system with and without a column.

Manufacturers have come up with other pulse dampeners to place in the flow from the pump to reduce this pulsation. The high-pressure versions are

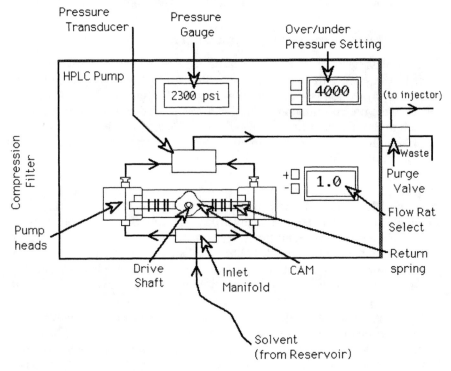

Figure 9.2. An HPLC pump.

metal cans with a long, tight coil of small-diameter stainless steel tubing coiled around itself. When a pulse comes down the line the coil of tubing flexes, then recoils through a spring effect and dampens some of the pulse. For some reason, manufacturers are often reluctant to admit that they have pulse dampeners on the pump. They describe them as "pressure filters" or "polishers." Generally, if your pump has a hole with two liquid lines leading into it, it is either a pressure sensor or a pulse damper.

Pump manufacturers continue to work to reduce pulsing. The first attempt was the two-headed pump, which is connected through an output manifold or pressure transducer (Fig. 9.3). One head delivers while the second refills. This

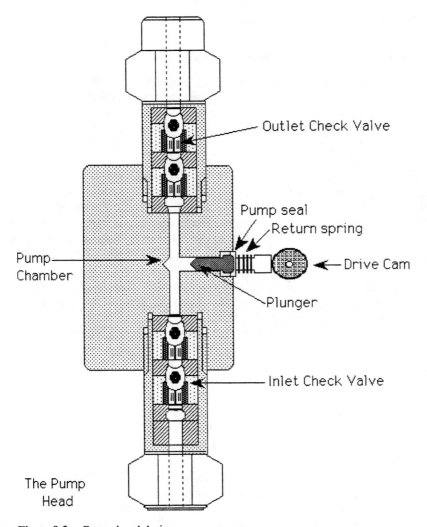

Figure 9.3. Pump head design.

improvement led to the first high-performance HPLC pumps, but it has an expensive downside: twice as many check valves, plungers, and seals, twice as much metal in the pump head, and more engineering knowhow leading to a higher cost. This eventually led to the three-headed pump. If this pump had been successful, we would probably be running V8s and in-line twelve pumps by now.

The next step was the electronically compensated pump. All pumps speed the motor as resistance increases to maintain a constant flow. These pumps also add a major motor speed-up during refill and repressurizations so that the pump spends 95% of its cycle in a delivery mode. With this modification, a pump with a single pump head and a pulse dampener could give 90% of the performance of a two-headed pump for 50% of the cost. This led to a price revolution for HPLC systems.

9.2.1 High- and Low-Pressure Mixing Controllers

Gradient systems are often oversold. They have two legitimate purposes. The first is for analysis of very complex mixtures with widely differing polarities. Some of this work results from the necessity to separate polar and nonpolar impurities resulting from the sample matrix. SFE cartridge columns for preinjection sample preparation can help to eliminate many of these gradient runs. The second purpose is for method scouting gradients that can be used to produce rapid running isocratic methods. Solvent gradients are used to move late running peaks off the column. To generate a solvent gradient you speed up one pump while slowing down the other, but still maintain a constant flow rate. Therefore, a gradient controller is, first and foremost, a pump flow controller. There are two types of gradient systems in HPLC: the high-pressure mixing gradient system and the low-pressure mixing gradient system. The first gradient systems were the high-pressure mixing ones. Two pumps were used to pump the individual solvents; once the solvents were pressurized, they were mixed before passing them to the injector. The remaining parts of the system were the same as in an isocratic system (Fig. 9.4).

The first high-pressure system I saw used a static mixer. The manifold that mixed the streams from the two pump heads on one pump had another inlet at 90° to the inlets for each pump head. You brought the line over from the other pump and screwed it into the manifold. You plugged a pump controller cable into the back of each pump from the system controller and you had a gradient system. More sophisticated systems brought both pump outputs together into a 50-μl magnetically stirred mixer and then out to the injector. In either case the mixing was done after the solvent had been pressurized in the pump head.

The problem with the high-pressure mixing system is that you have two very expensive pumps. You could actually benefit from having three solvents when you are doing method development, but that would mean adding even another expensive pump! There had to be a cheaper, better way of doing this. There is

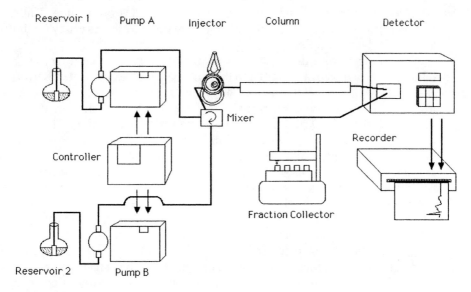

Figure 9.4.   High-pressure mixing gradient system.

a cheaper way, but not necessarily better. High-pressure, dynamically mixed gradient will always give the best gradient formation.

The second gradient alternative is the low-pressure mixing system (Fig. 9.5). This system uses a single pump, a gradient controller, and a solvent selection block fitted with solenoid valves. Each solvent line connects to the solvent block through a separate solenoid valve operated by the solvent programmer. Instead of controlling a pump's flow rate the controller switches each valve open for a specific length of time. If we dial in 80% A and 20% B, then valve A

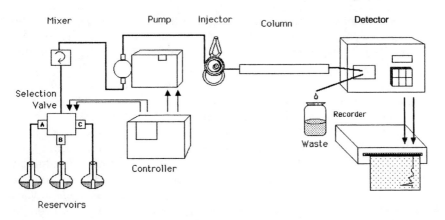

Figure 9.5.   Proportioning valve gradient system.

is open 80% of the time and the other valve is open 20% of the time. This solvent selection valve design is not limited to two valves; systems exist with three- and four-solvent selection valves.

The problems with this method of creating gradients is that it tends to put out plugs of each type of solvent and there is a heat of mixing that tends to drive gases out of solution before the solvent reaches the pump head. Plugs of solvent mean variability in the gradient unless a mixer is added; pumps are designed *not* to mix solvents. They have a first-in, first-out construction. The heat of mixing is a more serious problem. Gases released from solution end up as bubbles in the pump head and can and will lead to cavitation of the pump. Cavitation is what happens in your car when it vapor locks on a hot summer day. The HPLC pump locks up in the same way. It tries to pump solvent, but all that happens is that the bubble gets smaller, then bigger.

To overcome the first problem, that of mixing, manufacturers have introduced a variety of solutions. One provides a 2-ml mixing chamber after the pump head; this can be increased to 4 ml for critical gradients. That is a lot of dead volume to be added to the system before the column since it translates to time delay and imprecise gradients. A second manufacturer tried to add dynamic mixing immediately after the solvent selection block, but it just aggravated the degassing problem. A third manufacturer brought out a head and a half pump in which the second partial head piston is designed to mix the output from the first piston.

The degassing problem is generally solved by using helium-purged solvents. Solvents are purged with helium, sometimes under vacuum, and then run under helium sparging or under a helium demand valve. Using this technique, outgassing is prevented and these systems have become the most economical choice of the methods development laboratory, and, in many cases, the only choice considered for general research laboratories.

9.2.2 Checking Gradient Performance

In addition to the problem of heavy use of helium, there is some question about the efficiencies and reproducibility of the gradients they form. This question is easily answered. The best test of a gradient is from 0 to 5% and from 95 to 100%. In a high-pressure mixing system it is the point at which one of the two pumps is running the slowest, a true measure of pumping performance. In the low-pressure mixing system, it is the point at which the individual valve is open the least. Set up your system with MeOH in all solvent reservoirs and place 50 mM acetone in the A reservoir. Set your UV to 235 nm. Have your gradient programmer set for a binary gradient of 1%, 1 min steps from 0 to 10% B, 5% steps to 95% B, and 1% steps from 95 to 100% B, then reverse and repeat the same thing going down (1% steps from 100 to 95% B). Next, have it go through the same binary gradient sequence with A to C. If you have four solenoid valves, repeat with A to D (Fig. 9.6). You end up with series of step valleys if you trace the gradients. Ignore the gradient traces and inspect the recorder baseline. Since

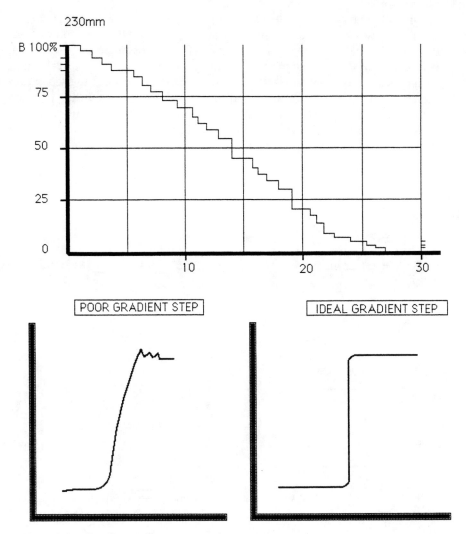

Figure 9.6.   Gradient performance testing.

acetone absorbs at 230 nm, you will see the actual gradients you produce. Ideally, each step should be vertical with sharp shoulders and no overshoot. In practice, you want to see only a bit of rounding at shoulders, a near vertical slope, and no ringing or overshoot at the new setting.

My preference today would be for a high-pressure mixing system if I had to run very complex mixtures on a routine basis. As a routine research instrument I would prefer a low-pressure, dynamically mixed system using a dual headed pump. I would use the deoxygenation apparatus (Fig. 6.4) to degas my solvents with helium and run them under a helium demand valve to conserve helium.

One other advantage of the two-pump gradient system is the ability to split it quickly into two isocratic systems by adding a second injector, column, and detector (Fig. 9.7). A majority of QA/QC HPLC runs made are fast running isocratic analysis once the proper operating conditions have been worked out. You could use two detectors in series and a gradient to scout the separation, then add an extra injector and column, hook both systems up to your two-channel integrator or strip chart, and run two separate isocratic runs at a time.

A two-pump system can be used for gradient scouting, separated into two isocratic running systems, then recombined for the next scouting problem.

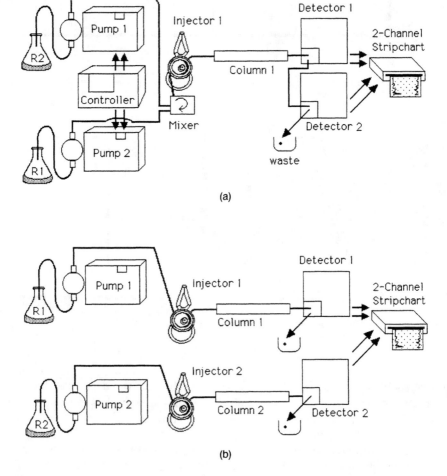

Figure 9.7.   Gradient to parallel isocratic systems. (a) Gradient system; (b) Parallel isocratic system.

This requires an extra injector and another detector, but is an excellent, cost-effective use of the HPLC dollar. It is actually like buying three systems in one: two isocratics and one gradient.

9.3 Injectors and Autosamplers

The next important component is the injector. Since it lies between the pump and the column, it also must be able to withstand high pressures. All injectors on the market work on the principle of the loop and valve. In the load position, system solvent is displaced from the injection loop with the sample at atmospheric pressure. Sample goes into the loop at the point closest to the column. Injection occurs by rotating the injector against a Teflon® seal to the inject position. This places the loop containing the sample in the solvent flow from the pump to the column. Since the solvent goes into the loop at the end opposite the sample, we achieve a last-in, first-out loading of the sample onto the column. No dilution of the sample occurs within the sample loop solvent volume. The loop is pressurized and the sample is washed onto the column (Fig. 9.8).

The injector is connected to the system immediately before the column head. Originally, injection was done by stopping the flow, then injecting through a septum. If you forget to stop the flow, syringe barrels explode or plungers fly across the room. Stop flow injections also present problems with run-to-run reproducibility and injected septum material onto the column head.

Autosamplers take this same loop and valve principle and automate the filling and handle turning sequence. The major differences between models on the market are in the way they get samples into the loop and the method of cleaning the needle between injections. Some models use air pressure or a piston to blow samples into the loop; others pull the sample into the loop with a vacuum or a syringe barrel. Most overfill the loop rather than try to do a partial injection. Various techniques are employed to wipe or wash the outside of the needle to prevent sample-to-sample contamination. There are elaborate systems to wash the inside and outside of the barrel. Other systems just wash the inside by pulling in some of the next sample and spitting it out to waste or back into the last vial.

A series of valves is actuated to actually get the sample into the column. Generally, low-pressure valves can be turned electrically, whereas high-pressure valves must be turned using air pressure. Most autosamplers require a source of compressed gas to run air-actuated valves.

9.4 Detectors

The detector controls the sensitivity with which compounds can be measured once separated on the column. To be effective, the detector must be capable of

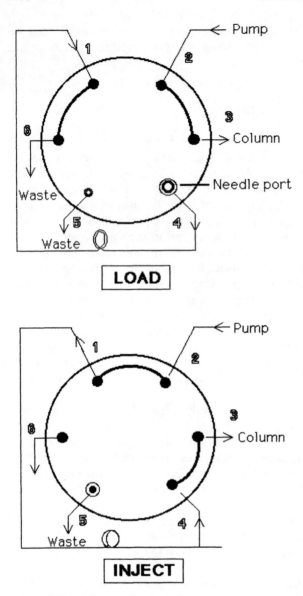

Figure 9.8. A loop and valve injector.

responding to concentration changes in all of the compounds of interest, with a sensitivity sufficient to measure the component present in the smallest concentration. Not all detectors will see every component separated by the column. Generally, the more sensitive the detector, the more specific it is and the more compounds it will miss. Detectors can be used in series to gain more information while maintaining sensitivity for the detection of minor components.

There are five main types of detectors used for HPLC: refractive index, ultraviolet, fluorescence, electrochemical, and conductivity. Infrared and mass spectrometric detectors have been used, but they suffer from solvent limitations. Many of these detectors are affected by temperature and recent sensitivity gains have been made by actively controlling flow cell temperature, by using "cold" infrared light sources, and by optimizing cell design to decrease gradient turbulence affects.

9.4.1 Mass Dependent: Refractive Index and Conductivity

The oldest detector, and the least sensitive, is the refractive index detector. Light from a source in the flow cell is directed first through the reference cell and then through the sample cell. The signal to the photodetectors is balanced with mobile phase passing through the sample cell (Fig. 9.9). The difference in refraction when the sample arrives causes the position of the beam to shift sending more light to one photodetector than to the other, resulting in a voltage shift in the detector output.

The detector of choice for preparative work, the refractive index detector can be used only in isocratic systems. It is very sensitive to turbulence, temperature, or solvent changes. Using a temperature-controlled compartment and a cold light source it is possible to push this detector to a 50–100-ng level of detection. It has the advantage of being a mass detector; the same weight of two compounds will give the same peak areas. However, make sure that the

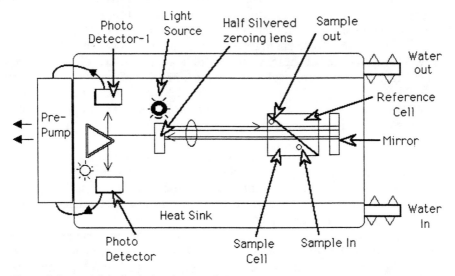

Figure 9.9. A refractive index detector light path.

compounds have refractive indices differing from the solvent. The flow cells on these detectors are fragile and will not tolerate backpressure. Simply blocking the flow from the detector for a few seconds will build up enough pressure to break a cell.

A second mass detector is the conductivity detector. It is designed to measure differences in conductivity in the flow cell against a reference electrode. Buffer in the mobile phase cuts the operating range and, therefore, the sensitivity. Gradient runs of either salt gradients or aqueous organic solvents cannot be tolerated by this detector. This detector is usually seen only in water analysis of inorganic cations and anions. To handle the problem seen with buffers it is usually run with a reverse osmosis column immediately before the detector, which helps to remove background buffer signal.

9.4.2 Absorptive: UV, Fluorescent, and Diode Array

The detector of choice for most separations is the UV/visible detector. These are available in two types: filter variable, fixed wavelength and fully wavelength variable. The fixed-wavelength detector uses only a single wavelength, although this may be changed with filters or by changing to lamps with different inherent wavelength output. The most common wavelength used is the 254 nm line of a mercury lamp, followed by 210 nm (Zn), 280 nm (Hg), and 329 nm (Cd).

A variable-wavelength detector, although more expensive than a fixed-wavelength detector, covers the whole wavelength range from 195 to 650 nm using a deuterium lamp as its source. Fixed-wavelength detectors within 15 nm of either side of their wavelength maximum give 4 to 10 times the sensitivity of the variable detectors. Fixed lamp life is 10 to 100 times longer and replacement cost is 5 to 10 times less than for the variable lamp. All of these differences—cost, sensitivity, and lamp life—have narrowed rapidly in the past 5 years. The variable-wavelength detector has become almost an HPLC necessity; it is the detector of choice in almost all laboratories. Light from the lamp passes through concentrating filters into the flow cell and out to be detected by the photodetector. In the variable detector, the concentrated light falls on a refraction grating, which splits the light into its individual components allowing selection of a specific wavelength to be passed through the flow cell (Fig. 9.10).

UV detectors are affected both by the mass of material present and its extinction coefficient at that wavelength. Some compounds will not absorb light at the wavelength used and will be missed. At present, these detectors can detect compounds, with good extinction coefficients, down to 100 pg. They probably could do better with purer solvents. Compounds with substituted aromatic chromophores usually absorb around 254 nm. Carbonyl compound and organic acids show "end absorption" at 220 nm; any solvent containing carbon–oxygen bonds absorbs too strongly below 220 nm to be used as part of a

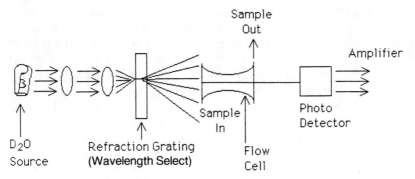

Figure 9.10. A variable ultraviolet detector light path.

gradient mobile phase. Carbohydrates are often detected at 190 nm, but dissolved oxygen in the mobile phase and oxygen in the optical path begin to absorb heavily and cut the sensitivity. UV detectors are also affected by temperature and solvent changes at very high sensitivity, but are reasonably unaffected at lower sensitivities. Flow cell design changes over the years have greatly reduced their temperature sensitivity. They offer the best detection for wide ranges of concentrations or different types of compounds of any of the commercial detectors.

The newest of the UV/visible detectors are the diode array models. The diode array detector modifies the position of the refraction grating placing it *after* the flow cell and adds an array of detection cells all looking continuously at a different wavelength (Fig. 9.11). The larger the array, the closer together these wavelengths can be selected. For any time point in the chromatogram, an absorption spectrum can be displayed from the array storage.

This may sound like the perfect detector, since almost all organic com-

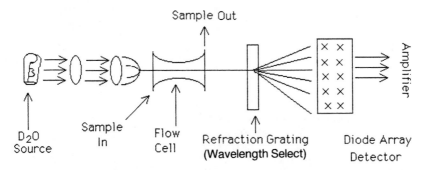

Figure 9.11. A diode array detector light path.

pounds appear to absorb light somewhere in the UV/visible region. It does suffer from a few problems. The most serious problem is *real time data display.* With all the information available, it can display only a maximum of two wavelengths at a time on a strip chart or an integrator or up to four to eight at a time on a screen. It becomes a fancy, very expensive variable UV detector. The mass of three-dimensional data (voltage–time–wavelength) output by the array can quickly overwhelm any but the most modern data storage systems. Extracting meaningful information from the mass of stored data from each run requires high-power computational and display technology (translation: high priced). A secondary problem is array chatter, which limits sensitivity. The third problem is the cost of the detector and associated data handling and storing computers. Once these problems are solved, the diode array offers great promise as a near universal UV detector.

The third type of absorption detector is the fluorescence detector. Fluorometers are more sensitive and more specific than UV detectors. To be detected, a compound must both absorb UV and fluoresce. Compounds that meet both criteria can generally be detected at 2 to 10 times more sensitivity by a fluorometer than by a UV detector. A fluorometer uses a variable UV lamp as an *excitation* source for the sample in the flow cell (Fig. 9.12). Light absorbed at one wavelength is promoted and emitted at a higher wavelength. The detector cell is placed at a 90° angle to the incident light and a cut off filter is used to remove light of the incident wavelength. Only the higher wavelength *emission* light escapes and is detected.

These detectors are often used to detect components of fluorescent derivatives prepared to increase the detection sensitivity of compounds with poor UV absorption. Both variable and filter variable, fixed-wavelength fluorometers are available for HPLC, with the same limits of lamp life and sensitivity seen in comparable UV detectors.

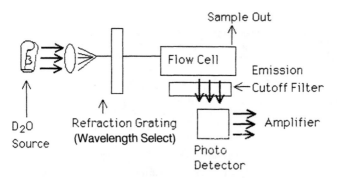

Figure 9.12.   A fluorescence detector light path.

9.4.3 Specifics: Electrochemical and Radiolabel

The most sensitive detectors in routine use are the electrochemical (EC) detectors. They also are the most specific detectors, seeing only compounds that are oxidized or reduced at the voltage applied across the flow cell. They need isocratic mobile phases that will carry ions and most separations are made with reverse-phase or ion-exchange columns in aqueous solvents. Current is applied across the flow cell from an operating electrode to a reference electrode. Control of the compounds to be oxidized (or reduced) is achieved by varying the applied voltage. Increasing the voltage potential increases the types of compounds that will be oxidized and detected. EC detectors have been used to detect 5 pg of rat brain catecholamines, probably close to the current operating sensitivity limit.

Occasionally, a laboratory will need an in-line detector of radiolabeled molecules. These detectors take the flow from the column or from an initial detector, mix it with fluorescing compound, and measure the fluorescence due to radioactive breakdown. A different system uses beads in the flow cell with an immobilized fluorescing compound, but these systems suffer from ghosting and cannot be used with very "hot" labeled compound because of secondary radiation problems. These systems are very useful with tritiated sample and less so with carbon-14. Some success has been reported with sulfur-32 label detection.

Detectors are not limited to solo use; they can be hooked in series to get more information from the same sample. In a serial operation, be sure that the refractive index detector or electrochemical detector is the last in the line. Their flow cells are more fragile than UV and fluorescence cells and will not take the increased backpressure. Keep the tubing diameter fine and as short as possible to avoid band spreading. You must correct for connecting tubing volume (time) delays in comparing chromatograms.

9.5 Fraction Collectors

Although not very common on laboratory HPLC systems, the fraction collector is an important part of preparative systems. It is an automated tray system that is designed to collect sample for a specific time period or collect a specific volume of sample, if it has either a drop counter attachment or can sense the pump flow rate. Some fraction collectors can be hooked to a detector and have a peak sensor attachment that allows them to collect only peaks while diverting the baseline to waste. It is important that the exit line from the last detector be equipped with a flow-through backpressure device so that sample can be passed to the fraction collector without too much band spreading or bubble formation in the flow cell.

9.6 Data Collection and Processing

Strip chart recorders, integrators, and computers are all means of storing and/
or processing information generated by the detector. When a compound is
detected in the flow cell, the detector sends out a signal with increasing and then
decreasing voltage. The recorder, running at a constant chart speed, records
this voltage change as a continuous trace versus time.

The integrator and the computer, using A/D converters, change the voltage
from a continuous, analog signal to a discrete digital signal. The integrator
sums the areas under each peak and stores the areas, peak heights, and peak
maxima time for each peak. A computer samples the voltage at preset time
points and records the sampling rate and each voltage displacement point.
Peak detection, integration, and peak identification are separate computer
operations, usually after run completion. Computer information processing
stores more information and requires more memory, but allows postrun display and reprocessing of the chromatographic information. Because of the low
cost of memory, many modern integrators have become dedicated microcom-
puters and allow reprocessing.

Strip chart recorders are the least expensive option, but areas and retention
times have to be manually calculated from the tracing. The more expensive
integrator, using small memories, gives us an annotated trace followed by a
report of areas (or peak heights) versus retention times. The computer requires
much more memory to store the one point per second (or more) required for
an HPLC run. However, it has much more flexibility in manipulation, display,
calculation, and report generation. Data processing will be covered in detail in
Chapter 14.

10

Troubleshooting and Optimization

Troubleshooting is the secret of keeping the wolf away from your laboratory door. Even if you do not want to work on the system yourself, knowing quickly how to find the problem can save you uncounted needless service calls and shorten the time the serviceman has to charge you for on any given call. Probably as much as 80% of the total HPLC problems are column problems, and 60% of these are due to bad water. If you are still using tap or triple distilled water, reading this chapter will simply be a waste of time.

We will start by reviewing the wetted surface from reservoir to flow cell output. I will discuss the minimum tools and spare parts you should have on hand and when to use them as well as some strange-sounding techniques to cut your solvent usage. A systematic approach to locating problems will be next. Then we will look at how to get the most out of hardware for data acquisition.

10.1 Hardware and Tools—System Pacification

The first step in the wetted path in the HPLC is the solvent reservoir holding our freshly filtered (and possibly degassed or deoxygenated) solvent. Most systems use a porous "stone" (5- to 30-μm filter) as a solvent line sinker. The tubing to the solvent inlet is wide-diameter Teflon®. In the solvent inlet line we may have another frit and a sapphire ball/stainless steel check valve. The wetted surfaces in the pumping chamber are all stainless steel except for the plunger and the seal. The plunger is made from sapphire (actually, it is beryl). It has good solvent resistance and strength along its long axis, but is fragile side to side. It will slowly accumulate a film from buffers that can be cleaned with toothpaste or light rubbing with scouring compound. The seal is a marvel of technology without which HPLC would be impossible. It is made of Teflon® containing a hardening agent, such as carbon fibers or ruby. Imbedded in it is a spring that squeezes the seal against the plunger when the pump head is tightened to the block. The plunger is lubricated by the mobile phase in the pump head allowing it to slide against the Teflon® seal. In the outlet line from the pumping chamber is another sapphire ball/stainless steel check valve. The compression fitting at the top of the check valve assembly is connected to 0.02-in. tubing leading toward the injector. All the tubing used is special acid washed, heavy wall HPLC tubing.

The pump outlet line usually next passes through a pressure sensor. One type is an in-line, double-coiled tube called a *Bordon tube.* As pressure increases, the tube begins to straighten, blocking light falling on a sensor, which is translated into a meter deflection; the more pressure, the more deflection. From the sensor, the tubing travels on to a flush valve allowing venting through a 0.04-in. tube to the atmosphere for rapid exchange of solvents. With the valve shut, the flow next goes to the injector.

The injector's wetted surfaces are stainless steel and acid washed tubing, except for the rotor seal. The seal is a block of impregnated Teflon®, like the pump seal, drilled out for the loop and bypass pathways. Movement from inject to load is lubricated by the Teflon®. The injector outlet is drilled to 0.01 in. and is equipped with a compression fitting carrying tubing of the same diameter.

The next step in the pathway is the column. The compression fitting on the inlet endcap leads to the stainless steel frit at the top of the column. The column itself is a heavy-walled stainless steel tube filled with packing and mobile phase. The outlet end is identical to the inlet. Moving on down the wetted surface, we find 0.01-in. tubing leading to the detector flow cell.

The detector inlet usually passes through a coiled stainless steel tubing heat exchanger and into the flow cell. The flow cell is the most complicated part of the system. The body is stainless steel or quartz, the windows are quartz, and if it can be taken apart for cleaning, there is usually a Teflon® gasket between the stainless steel body and the quartz window. Finally, we move out of the flow cell into the narrow-diameter Teflon® tubing of the outlet tubing and into a backpressure device in the waste vessel.

The purpose of this tour was to review the solvent path from the solvent's point of view. I also wanted to point out that the system is made from resistant material, except for the column.

Wear points (and resulting problems) are as follows:

1. pump: check valves (buffer crystallization), seal (dry tearing leading to leakage), and plunger (breakage);
2. injector: rotor seal (tearing—sample carryover) and needle seal (scoring—leakage);

3. column: (particulates, precipitation in bed, fines);
4. detector: flow cell (breakage and hazing) and lamp (aging).

Most of the pump problems come from buffer precipitation in going to immiscible solvents; the rest come from normal wear. Plunger breakage may come from buffer accumulation, but is usually the result of removing the pump head without drawing the plunger all the way back. Injector problems come from particulate scouring from buffer crystals and unfiltered samples. Detector problems come from buffer blockage of the outlet tube, outlet tube restrictions (such as fingers held on the end), and sample decomposition.

Realizing that 80% of all system problems are column problems, it becomes clear from the preceding that the next major cause of a problem is "doing something dumb." Buffer use will usually cut seal life in half. Generally, crystallization is caused by failing to go through an intermediate solvent. This leads to buffer crystals throughout the system, which causes wear on any moving surface and plug lines.

Since prevention is the answer, but is often ineffective, how do we deal with these problems? The first step is to get the column out of the system. Second, to treat the whole system we need to replace the column with something to provide a liquid path. Most systems are designed to operate best under some backpressure. I generally replace the column with a column blank. This is easily made from tubing, unions, and compression fittings. I have found that 1 ft of 0.01-in. tubing equals 50 psi backpressure at 1 ml/min of acetonitrile. Please check this since the tightening of compression fittings will vary and can change backpressure. With compression fittings and unions, you should be able to match the ends of your column exactly.

With the column blank in place, you can now proceed to check out the mechanical system unhindered by overriding column effects. Check valve problems can usually be seen on the strip chart recorder magnified by the low backpressure. Injection of a standard into the system should give an instant response. This signal can be used to diagnosis detector, injector, and recorder problems. I believe that a column blank belongs in every HPLC tool bag. It is an ideal tool for system washing with harsh reagents, which would seriously damage the column.

With the column blank in place, we are ready to do a general system cleaning. The technique is called *pacification*. It involves washing the system with 20% nitric acid. It sounds harsh and is, but your system should be resistant to it. Check your manual to ensure that all wetted surfaces are stainless steel, ruby, Teflon®, and quartz. If you cannot tell from the manual, call the manufacturer. Use pacification only with UV detectors; I cannot guarantee others will be resistant.

Before we proceed, make sure you *remove the column*. Trying to do pacification with a column in the system will, at the least, ruin the column. After removing the column and inserting the column blank, wash with water for 15 min at 2 ml/min to remove any buffer. Follow this with 20% HNO_3 (6 N) for

15 min at 2 ml/min, then wash out with water until neutral pH is reached. This last step takes a long time.

I usually recommend pacification be done on the last Friday of the month on a routine basis. Starting late in the afternoon, you wash with water, *remove the column,* replace with the column bridge, wash with acid and then water, then let the system run over the weekend at a slow (0.1 ml/min) flow rate. (Leave at least 600 ml in the flask when you leave Friday.) On Monday check pH, replace the column, equilibrate with mobile phase, and run your standards.

Pacification *after removing the column* is an excellent cleaning technique for the whole system. It tends to remove buffer precipitation from check valves, organics from the injector rotor, and deposits on detector flow cell windows. This treatment is recommended to protect the system from halide attack on stainless steel. I have been told by experienced protein people that pacification will protect a system against 200 mM NaCl for a month. This assumes you wash out the NaCl before you shut the system down.

Again, please check your owner's manual or talk to the manufacturer of your system before attempting pacification. All the systems I have worked with have been resistant to this treatment. They are built from stainless steel, Teflon®, sapphire, or ruby, all of which are resistant to this concentration of nitric acid for short periods of time.

However, there are some machines that may not be able to stand the treatment. Finally, I would like to remind you one more time to *remove the column before attempting to pacify your system.* These reminders may seem excessive, but I had one student pacify a C_{18} column after being warned six times in class. Please do not make the same mistake!

Other tools you will need for HPLC are fairly standard for a laboratory, with a few exceptions. I keep two of the small open wrenches supplied with an HPLC system, a $\frac{1}{4}$- and a $\frac{5}{8}$-in. box wrench, a 6-in. crescent wrench for column work, a file, a Terry tool tubing cutter, a couple of pairs of blunt nose pliers for tube cutting, a reversible screw driver, a jeweler's screw driver for phoenix blocks screws, a universal allen wrench set, and a dental mirror (for seeing behind boxes to hook up leads). The key is to lock up your tool kit to prevent tool evaporation.

Spare parts that are needed are compression fittings and ferrules, plunger and injector rotor seals, an extra plunger, column frits, and injector needle port seals. If you do not use pacification, you might want to keep a set of check valves on hand. I always have one coil each of 0.01- and 0.02-in. tubing in addition to my column blank. A replacement solvent inlet line with a porous stone is useful in case of corrosion. If you filter solvents, you need both cellulose, nylon, and Teflon® filters. You also need a back-up lamp for your detector.

If it takes a while to get replacements, double the amounts of the preceding parts and add a detector flow cell, another C_{18} column, and a full pump head. If you are going to Antarctica for the season, an extra injector, pump, detector, strip chart recorder, and a case each of strip chart paper and pens might be nice.

One of my customers found that his back-order time in Little America was 14 months.

10.2 Reverse Order Diagnosis

The most commonly reported symptom of a bad system is baseline drift; the second is a noisy baseline. Baselines that drift up and down are almost always the result of peaks coming off the column. Baselines that drift up continuously can be caused by garbage coming off your guard column, a bad lamp, or decomposition on the flow cell window. Noise can come from almost anywhere in the system.

When you have a problem, do not guess as to where it might be. Start looking for problems from the strip chart end of the system. Prove that the strip chart is good, then use it to check the detector, and use the detector to check the injector and finally the pump. Using this systematic, reverse-order analysis will save you time and frustration.

As we said in the column section, the first step is to remove the column and see if the problem goes away using a column blank. Pressure problems can be determined by looking at the pump pressure. Other problems can be diagnosed from the detector digital display and the recorder baseline. Of course, with the column gone, there is no connection to the detector. Here is where the column blank shows its versatility. When we remove the column, we replace it with the blank. Now we can run the pump and make injections and see the effects on the detector/strip chart. The 5 feet of tubing allows the pump to run with some backpressure, but not as much as the column; pumping problems are magnified and injections fly through the detector.

The strip chart is the most ignored part of the system, but it can contribute its share of problems. People will usually spend $20,000 for an HPLC, then grab some old strip chart off the shelf, blow off the dust, hook it up, and wonder why things look terrible. I had a customer who did exactly that, then called complaining that his pumps were varying by 10% in flow rate and his chromatograms were not reproducible. I went into his laboratory, disconnected the HPLC, and timed the strip chart bed speed at 0.5 cm/min. We found it varying by 10% on either side of this speed. It had been used at 2–5 cm/min on a GC for years and somewhere a drive spring was overstretched.

There are two modules in a strip chart: the electronics and the mechanical. The trick is to look at one piece at a time. Disconnect the detector leads, turn off the bed, and watch the pen. Does it sit quietly or chatter up and down? Noise at this point comes from the strip chart electronics. If it is quiet, turn on the bed and let the pen trace; is it a flat baseline? If so, short across the leads and make sure the pen deflects without sticking. (This would show up as a plateau in a chromatogram.) Lubrication or drive wire replacement would fix the problem. Do not get much oil on the slide part; it just traps dirt. Spray some WD40 on a Kimwipe. Wipe the bar, then wipe off the excess. Next, time the bed. Is it accurate at 0.5 cm/min where you will be using it? If it passes these tests, we are ready to hook up the detector leads and move on.

The detector also has two modules: the electronic, including the lamp, and the flow cell. Flush the flow cell with strong solvent, then turn off the flow and observe the strip chart signal. An old lamp and a noisy baseline might indicate a lamp replacement is needed. Tungsten lamps used in fixed-wavelength UV detectors are good for 1000 to 5000 hr. Variable lamps have a 1000-hr meter, but may get less depending on the wavelength. Low wavelengths may lose sensitivity faster than wavelengths around 254 nm. If you change the lamp and the baseline stays noisy, suspect the electronics. If the baseline continuously rises, you may have a compound coated on the window that is decomposing under UV excitation. Some older model UV detectors, like refractive index detectors, are very temperature sensitive and the baseline will follow air conditioner cycles or drafts. The baseline effects are fairly long and you may have to shield the detector or thermostat it.

If the detector passes the static test, turn on the flow and watch the baseline. A noisy baseline at this point is probably coming from before the detector. Realize that a reciprocating pump is noisy and it now lacks pulse dampening from the column. While we are here shoot a sample; the response should be instantaneous, straight up and down. I have not tried it, but you might be able to quantitate lamp strength by shooting a known standard over a time period and measuring the deflection.

The next component to check is the injector. Usually, only three things happen: leaks and plugs, which are immediately obvious, loop contamination, and carryover. Plugging can be localized by working backward from the column connection port. Put the injector first in the *inject,* then in the *load* position with the pump running until the pressure drops. If the problem is in the loop, reverse the loop and use pump pressure to blow out the plug.

Once flow is cleared shoot the injector; that is, just throw the handle into the inject position. Do you see a small positive peak? Next shoot a normal injection size of a strong solvent; does that give a peak? If both are yes, you may have a sample loop that needs cleaning. Next shoot a sample of a good UV absorber, then shoot the injector without putting in a new sample. Do you get a "volunteer peak" without shooting a new sample? This ghost peak can be caused by scratching of the injector rotor causing sample carryover, indicting that the rotor seal needs to be replaced.

The last module to be inspected is the pump. Leaking is the most obvious problem. There is always some flow around the plunger, but this evaporates with solvents. With buffers, this leads to a weeping, which is messy, but also harmless if washed off periodically. When the seal deteriorates, solvent comes out rapidly and pressure cannot be maintained. Stop the pump with the plunger back, loosen the screws holding the head in place, and carefully slide the head off the plunger. Dig out the old seal, clean the plunger with mild abrasive if needed, and replace the seal and the pump head.

If the pump does not deliver solvent, do not immediately assume that the plunger is broken. Open the compression fitting at the top of the check valve to release pressure. Pressurize the inlet line with a syringe full of solvent and turn the pump on. Often an air bubble will cause the pump to cavitate and stall. This is often a problem with pumps having filters in the inlet check valve. This technique should cause air bubbles to come out of the outlet fitting. Simply reseal and go back to work.

If the head does not leak and is delivering solvent, catch solvent in a graduated cylinder and time the flow. If you get much less than the calibrated flow, your outlet check valve may be dirty, allowing back flow. Watch the surface of your reservoir; if it rises and falls repeatedly during pumping and you can get pumping pressure you may have a dirty inlet check valve. Both of these problems can be treated with pacification if caught early enough.

You now have the tools to find most of the problems in the system. Remember that 80% of the problems are column problems and about 60% of these come from dirty water. If the problem fails to resolve with the column, put in a column blank and do reverse order diagnosis. It will be amazing how much time you save.

10.3 Introduction to Data Acquisition

Inexpensive integrators have made life easier in one sense, but have complicated it in another. The computer extends this dichotomy. I will discuss briefly the variables that must be controlled in an integrator and how to use it in a research versus a clinical environment. I will also explain when and why you might want to hook up to a computer.

The output from a detector is a voltage that varies with changes in concentration and the compounds' extinction coefficients. In a strip chart recorder, this voltage change is plotted as a vertical displacement versus time at a constant chart speed. We can measure a peak area from the vertical maximum and the peak width. The integrator plots the same output, but time-stamps the peak maxima. Internally, for each peak, it stores in memory two numbers: the peak area (summed from all the points from leaving baseline to finding it again) and the maxima. At the end of the chromatogram, it reports these areas versus time, sums all areas, and reports each as a percentage of the total.

The computer handles data differently. It measures the voltage at a preset time interval and stores each displacement as a digital word, usually at a rate of 10–20 points/sec. The computer requires much more memory than the integrator to store a single chromatogram; however, it can use these raw data to recalculate, redisplay, and compare chromatograms.

If you are using the integrator for scouting or research runs, where run-to-run reproducibility is not critical, it is very easy to initialize conditions. Most integrators have a slope test or calibrate button. When you push the button, the integrator looks at the baseline for about a minute and sets all variables. The only time you need to recalibrate is when you change the detector sensitivity.

Integration is started either by a signal from the injector or by a time programmable delay trigger by the injector or the start button.

When you are in a clinical environment where reproducibility is critical, you need to know and understand the variables that the calibration sets. An integrator should be able to integrate close neighbors about four times more accurately than you can by hand. The integrator sets three variable levels: peak width, slope rejection, and noise rejections. It also makes decisions on how to integrate unresolved peaks. As peaks widen (or narrow) in later parts of the chromatogram, the integrator doubles (or halves) the peak width value.

To achieve maximum reproducibility you must make these decisions instead of leaving them to the machine. You will have to do a methods development project. Turn off the automatics, shoot a sample, adjust a variable, then repeat until you have it right. When setting variables manually, we first set the peak width followed by slope rejection and then noise rejection. If peak width is correct, the print gap, left when the integrator prints the retention time, will fall half way down the back side of the peak. If it falls closer to the top or on the front side, the peak width is too small. Normally, 5 msec is the preset time and will handle fast eluting peaks. The integrator is designed to automatically double this setting as late running peaks broaden by diffusion. You can usually override this function and program timed peak doubling, so each chromatogram is handled the same.

When the machine looks at the baseline it finds the largest peak and uses its front slope to set slope rejection. Any slope greater than that is a peak and is time stamped and integrated. Once the slope detect is set the integrator averages the heights of peak top level and uses this to set noise rejection. If slope rejection is too high, the integration start and stop marks will fall on the peak shoulders. If noise rejection is too low, you get an almost continuous printout of retention times even though peaks may not be seen.

Unresolved peaks pose a problem for the integrators. Depending on their size and the baseline slope, the integrator may do a vertical drop to baseline from the minimum between peaks, do tangential skim, or draw the baseline valley to valley and integrate. It may do it differently from run to run. You can make a decision on each peak pair, and build a time program forcing the same type of integration each time.

The basic difference between the integrator and the computer is that the latter stores more information and can be more flexible with how it handles it. With all the points that make up the chromatogram stored, the computer, with the right software, can regenerate all or part of the run, blow it up or shrink it down, leave out extraneous peaks, or do a point-by-point comparison with other curves. Flexibility adds complications; setting up and using the computer are more complicated and costly than an integrator.

At the start of the chapter, I promised to describe a technique for solvent use reduction. A customer in a government laboratory cut his solvent use by a factor of 5. He started with 5 L of fresh solvent, pumped it through the column, made his injection, and recorded his chromatogram while recycling his waste liquid to his stirred solvent reservoir. Before each new injection, he autozeroed his detector baseline. If you have an injection-sensing switch, it often can be used to autozero the detector automatically. After pumping a total of 25 L through the column, he discarded his solvent and started again with a fresh batch.

This sounds very unsanitary, but it works very well. It can only be used for isocratic runs, not for gradients. Injection samples (nanograms per microliter) introduced into the 5-L reservoir only slowly increase the baseline absorption levels. Nothing passing through the column once will stick on a subsequent passage.

Part III
HPLC Utilization

CHAPTER
11

Preparative Chromatography

The preparative use of HPLC is often overlooked in the rush to analyze. There are two outputs from a detector: the electrical signal to the strip chart and the liquid output. Most columns and detectors are nondestructive; aqueous samples can be run directly without derivatization. The same system used for analysis can be used for all but the largest preparative runs. Usually, all that is required is a change of column and a slight modification of the running conditions to accommodate the increased sample concentration (Table 11.1).

Speed, load, and resolution are the three tradeoff considerations that must be balanced to optimize the three levels of preparative runs (Fig. 11.1). Analytical preparative is concerned with isolation of up to microgram quantities of material and with obtaining enough material for spectrometric analysis; the most important factors are speed and resolution.

As we move to semipreparative separation, in the milligram range, we are usually purifying analytical standards or recovering impurities to do trace compound analysis. Resolution is still very important, but now load, not speed, is the tradeoff. "True" preparative at the gram level can be running using a semipreparative column in an analytical system by making multiple injections and collecting similar fractions. It is usually run, however, in a true preparative HPLC system with much higher flow rates. Load is the major tradeoff; speed is secondary, with resolution the least important. Here we are gathering grams of material for biological testing and structure analysis, and as reaction intermediates.

Table 11.1   A Guide to Preparative Scale Up

	Analytical	Semipreparative	Preparative
Size	4.6 mm × 25 cm	10 mm × 25 cm	25 mm × 25 cm
Packing	3, 5, 10 $\mu\mu$	5 μm	40 μm
Max. Load	150 μg	100 Mg	5 g
Flow Rate	1 ml/min	4.5 ml/min	30 ml/min
Resolution	1	0.8	0.3
User	Clinical analytical	EPA/process	Standards organic synthesis

11.1 Analytical Preparative

These separations are run at 1.0 ml/min on the same 4.5-mm × 25-cm column used for analytical runs. Normally, in analysis we shoot from picogram to nanogram quantities. Most separations maintain their resolution until we reach about 1 μg. The valleys between peaks then begin to rise, indicating some overlap.

If we increase our first peak k' to 8–10, we can increase the interpeak gap allowing us to load to about 10 μg of compound/injection. As we increase the amount of sample, we need to go to lower detector sensitivity. We can increase flow rate to 2.0 ml/min, but we will lose resolution by doing so. Generally, we have no problem increasing sample concentration and keeping the same injec-

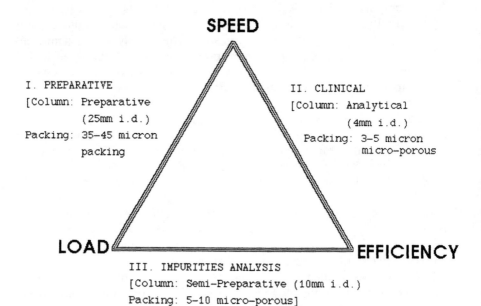

Figure 11.1.   Speed–load–resolution decision.

tion size. If necessary we can increase to the next size larger loop without affecting resolution.

If we have to increase load higher, say to obtain a 50 μg sample for NMR, we can use the shave/recycle technique to be described in the "True" preparative section. These runs must be made isocratic and column overload occurs at 100–150 μg for most compounds. If this much material is needed, it is better to switch to a semipreparative column, which can easily produce milligram quantities in a single pass.

11.2 Semipreparative

Semipreparative separations are made on a 10-mm $\times$ 25-cm column packed with the same 5- or 10-μm packing used in the analytical separation. Simply replace the column and equilibrate with the analytical mobile phase. A 1–5-mg sample can be injected with a flow rate FR_2 calculated from the following formula:

$$FR_2 = FR_1 \times (D_2/D_1)^2$$

where FR_1 is the analytical flow rate, D_2 is the semipreparative column diameter, D_1 is the diameter of the equivalent analytical column, and is the square of the diameter difference. Here we would use a flow rate of about 5 ml/min.

By using techniques to increase k', we can push the load to 20 mg. Going isocratic and using shave/recycle the load can be increased to 100 mg with column overload occurring at 200–300 mg.

11.3 "True" Preparative

Preparative separations in the grams per injection level are different. Separations are run isocratic in 1- to 3-in. columns packed with large pore packings (35–60 μm). An analytical, two-pump system can just barely reach the 20 ml/min flow rates needed to run a 1-in. column. Special preparative HPLC systems deliver flow rates of 50–500 ml/min to handle the larger bore columns. A stream splitter is used to send part of the flow through a refractive index detector designed for concentrated solutions.

Injection samples need to be as concentrated as possible and this leads to problems. A column acts as a sample concentrator. If the solution starts out saturated, it will supersaturate on the column, precipitate, and plug the column. I have seen a column with a 3-cm-deep plug that had to be bored out with a spatula. A couple of injector loops full of the stronger solvent will clear this, if there is still some flow, but the separation will have to be repeated. It is better to dissolve the compound, then add a half volume of solvent, ensuring that there will be no precipitation.

A technique called *shave/recycle,* mentioned earlier, allows separation of a pair of close peaks. To use shave/recycle it is necessary to plumb the HPLC system so that the output from the detector can be returned to the HPLC pump through small-diameter tubing and switching valves (Fig. 11.2). Twenty-thousandths tubing is used to connect the detector output to valve 2, the waste recycle valve, 0.02-in. tubing connects from valve 2 to valve 1, the solvent select valve, and finally a third valve 3, the collect valve, can be placed in the waste line from valve 2.

The analytical separation is used as a guide to selecting load conditions. First, k' (Fig. 11.3a) is increased until the first peak comes off with a $k' = 10$ (Fig. 11.3b). Load is scaled up until the valley of the two peaks is just visible (Fig. 11.3c). If there are peaks running later than our pair, we will have to inject the sample and collect the fraction containing the compounds of interest for injection. If the only impurities come off before the compounds of interest, they can be removed after making the injection. The preparative instrument can be a little intimidating because things happen so fast. With a top flow rate of 500 ml/min a liter flask is effectively filled in less than 2 min. The first time you run the analysis, I would suggest using the slowest flow rate possible.

To begin a run (Fig. 11.3d), a sample must first be injected with valve 1 turned to reservoir and valves 2 and 3 to waste. You can use either a very large loop and valve injector or a stop flow injection in which the sample is pushed into the solvent line through a injection port, or you can pump the sample in using either the main HPLC pump or an analytical, loading pump plumbed in through a three-way valve 1. After injection (Fig. 11.3d), you will see the void volume peak followed by any early running impurities, which are all run to waste. If you are using a loop and valve injector, wash the loop with 6 loop volumes of solvent, then *turn the handle back into load.* This is important because it removes a major source of dead volume from the recycle pathway.

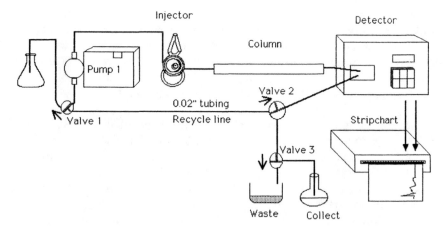

Figure 11.2. Recycle system.

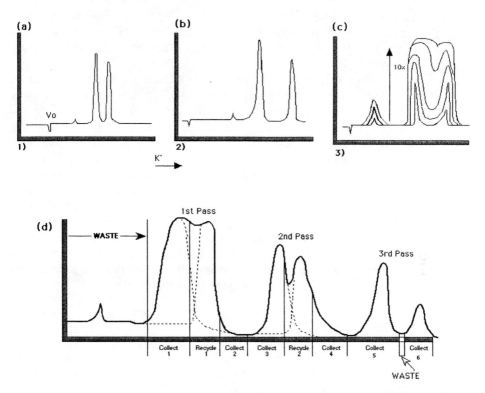

Figure 11.3. Shave recycle. (a) Analytical; (b) k' increase; (c) Overload; (d) Preparative: shave/recycle.

As your first peak begins to elute, switch to collection of fraction 1 by turning valve 3 to collect. Continue to collect until you are over the maxima of peak A and one-third of the way down to the valley between the peaks. (Peak A tails badly into B, but there is little of B in A until we get well into the rear slope. Be brave, you can always inject.)

At this point switch to recycle by turning valves 1 and 2 at the same time. We want to send the contaminated portion back through the pump to the head of the column for further separation. We continue to recycle until the detector shows we are well down on the backside of peak B. (Remember, A is tailing into B.). Shut recycle valves 1 and 2 and switch to collection of fraction 2 in a clean flask. Stop collecting 2 when we reach the baseline by turning valve 3 to waste and changing to a clean collection flask.

Very quickly, peak A should begin emerging from its second pass through the column. Open valve 3 and collect fraction 3 in the clean flask. We will continue the cycle, (1) collect peak A, (2) recycle the middle by opening valves 1 and 2, then (3) close valves 1 and 2 and collect peak B, until we have separated the peaks (three passes is usually enough). The slow flow rate is fine for beginners, but you will find yourself quickly using only the maximum flow rate of the system. Have plenty of volunteers on hand. People will be rushing around with flasks of collected sample trying to get to a flash evaporator before sample start to crystallize.

Once all fractions are collected, we can take them to our analytical apparatus to ensure purity, then combine odd fractions—1, 3, 5, and so on—for recovery of peak A, and even fractions—2, 4, 6, and so on—for recovery of peak B. Volatile, organic solvents can be rotary evaporated for sample and solvent recovery. Evaporation of large volumes of water mixtures from samples eluted from reverse-phase columns can be time consuming. You can use the preparative HPLC to recover pure compounds from aqueous solution. Dilute the combined fractions for peak A 5- or 10-fold with water and pump them back onto the column, either through the injector, the injection pump, or the main system pump. This increases the compound's k' causing it to be retained at the column head. Then immediately elute the compound with a strong, volatile solvent such as methanol. Your sample can be recovered by rotary evaporation from the small volume of strong solvent needed for elution. Each purified compound can be recovered in turn using the same technique.

12

Sample Preparation and Methods Development

12.1 Sample Preparation

Sample preparation is the key to getting the most out of an HPLC system. Unless you are working with purified standards or examining a compound in a very pure matrix, your chromatography becomes complicated with extraneous compounds.

A biological matrix such as plasma provides a good example. Nature generally tends to make metabolites of ingested materials more polar than the original compound. These polars exist to aid in elimination and excretion as well as to serve as building blocks and reaction components. At the same time, nonpolar molecules are present in transport and structural roles and end up in the circulating blood.

The effect on chromatography is to complicate the separation greatly. If we consider a reverse-phase separation, the first thing we notice on injection of a plasma sample is an almost irreversible binding of protein to the column. Even after protein removal, we find polar peaks, which overload the early part of the chromatogram and tail into the compounds of interest. The components more nonpolar than our compounds of interest adhere to the column and must be washed off before the next injection. To ensure polar elution before our target compounds and nonpolar removal afterward, we are almost forced to run solvent gradients.

Sample preparation techniques are aimed at removal of as many of these extraneous materials as possible before injection onto the column. The expected result should be a dramatic reduction in run times, hopefully to a fast running isocratic separation instead of the gradient. A side benefit of much of this sample preparation is often a trace enrichment, an increase in sample concentration with a corresponding increase in detectability.

The first step in preparing a sample for injection is to ensure that it is completely dissolved and to remove particulate matter. If we are working with blood samples, we need to get rid of blood cells, which is done by centrifugation. Be aware that in removing both particulates and red blood cells, there is a chance that you may remove some of your compound of interest. This needs to be checked by adding known amounts of your target compounds, removing the interfering materials, and then quantitatively checking your recovery of targets.

12.1.1 Deproteination

The next general step is to remove charged molecules that interact with silica, which assumes, of course, that you are not trying to recover proteins from solution. Since silicic acid is a weak cationic ion exchanger, the compounds we are trying to remove will be positively charged. In serum, the most common of these are proteins. Proteins also can interact with bonded-phase columns through nonpolar partitioning. It is usually best to avoid putting them on the column since removal is difficult and time consuming.

As mentioned earlier, proteins can be removed by ultrafiltration through a very fine membrane filter. Ultracentrifugation at high speeds can also be used to separate proteins from smaller molecules based on size differences. The most commonly used protein removal techniques for HPLC involve protein denaturation. Most proteins are denatured by heating. If the compounds to be analyzed are temperature resistant, blood cells can be spun down and the crude mixture remaining can be boiled and then filtered or centrifuged. Particulates and denatured protein are removed together.

Chemical denaturation of proteins tends to be more efficient and less harmful to sensitive compounds. Acidification with trichloracetic acid (TCA) (5% in final solution), centrifugation to remove protein, and neutralization with sodium hydroxide remove better than 99% of the protein. A second reagent used for protein precipitation is perchloracetic acid. After protein precipitation occurs, excess perchloracetic acid is precipitated as $KClO_4$ by neutralization with potassium hydroxide. Both of these acid treatments, however, suffer from problems. TCA absorbs strongly below 230 nm eliminating the use of low-wavelength detection. The perchloracetic acid treatment leaves large amounts of salt in solution, which can precipitate with organic solvents or cause major early refractive index upsets of the UV baseline.

Probably the best chemical precipitant is acetonitrile. Acetonitrile has the advantages of being a common solvent for HPLC and of being UV transparent to 190 nm. Mixing and centrifugation of an equal volume of sample solution and acetonitrile will lead to precipitation of about 95% of the proteins. Nonpolar proteins, such as the albumin fraction, remain in the sample. The supernatant can be injected directly if a guard column is used to remove the last 5%

of the protein. The guard column will need to be repacked or inverted and washed into a beaker with 70% acetonitrile/water containing 0.1% trifluoro-acetic acid periodically to prevent protein breakthrough to the main column.

12.1.2 Extraction and Concentration

The next step is to obtain the compounds of interest free of interferences. This has traditionally been done by extracting only those compounds from the solution leaving all others behind, an ideal, but seldom realized goal.

Extraction of nonpolar compounds using equal volumes of sample and the Folsch mixture (2:1, chloroform/MeOH) gives a very broad polarity cut. Everything from steroids to triglycerides are pulled down into the chloroform-rich bottom layer. Extraction with methylene chloride from a sample acidified with sulfuric acid is more specific, pulling in steroids, phospholipids, fat-soluble vitamins, and free fatty acids. The triglyceride fraction can be extracted using i-PrOH/hexane (1:9) with little emulsification.

After extraction, these fractions should be dried to remove water. When dry, the extraction solvent is removed by evaporation and the sample is reconstituted with a solvent or mobile phase before injection. Care must be taken that these evaporated samples go completely back into solution. Sonicating the sample with your starting mobile phase is usually sufficient. However, at least the first time you perform an extraction, it is always good technique to sonicate the dry down tube with a strong solvent and inject this wash as a check that everything redissolved. For gradient work, the stronger of the two mobile phase solutions is an excellent choice for this second sonication solvent.

It is always a good idea to make sure particulates are removed from these sonicates, or for that matter from any sample. Using centrifugation or filtration as a last step before injection protects the column filter from plugging and the system from pressure buildup.

12.1.3 SFE (Cartridge Column) Preparations

One of the most useful additions to the chromatographer's armament has been the off-line precolumn, sometimes referred to as SFE, sample preparation and extraction, columns. These disposable, low-pressure cartridge columns contain large particle packings with the same bonded-phase as the HPLC column. The theory is that anything that will go through them will go through the HPLC column. Because they are silica based, protein sticks to them very tightly. The 50% acetonitrile solution prepared earlier can be passed down one of these cartridge columns to remove the last traces of protein.

Cartridge columns consist of 0.5–1 g of 40-μm particles sandwiched between 30-μm frits. The column's body may be a tube shrunk around a sandwich of bed support frits and packing. In another type, the frits and packing may be pushed into a small syringe barrel. The 0.5-g SFE has a sample capacity of about 25 mg of organic compounds and a void volume of about 1.5 ml.

The SFE can also be used for extractions. Reverse-phase cartridges are available that, after activation with 2 ml of methanol or acetonitrile followed by 2 ml of water, can be used to remove nonpolar materials from solution. This is true chromatography, not filtration. The adhering material can be eluted in step gradient fashion with increasing nonpolar solvent fractions. Although they do not have the efficiency of an HPLC column, they can separate classes of material.

For partition extraction, it is important that the compound is in its uncharged form. Extraction of organic acids from aqueous media is easier if you first acidify with sulfuric acid. Bases are easier to extract above pH 11. Most compounds are more volatile in their uncharged forms, so be careful during removal of the extraction solvent. The C_2 to C_4 organic acids are often lost during evaporation of methylene chloride extracts under a nitrogen needle dryer. If you plan to derivatize for increased sensitivity, add the coupling base to the methylene chloride solution before evaporation. The organic acid salt formed is much less volatile and will stay in the extraction tube.

A technique that works well in extraction of charged molecules is the use of ion pairing reagents. These counterion compounds are used in the chromatography to make charged polar compounds appear more nonpolar so that they will adhere to a C_{18} column. The same technique will work when extracting with organic solvents or with nonpolar cartridge columns. A word of warning, however, is needed: Unless you are planning to run ion pair chromatography on the compounds in your main HPLC column, you may find the ion pairing reagent difficult to separate from the target compound. Generally, if you must remove the ion pairing reagent, pick the most nonpolar ion pair reagent available, extract, neutralize the target compounds charge, and back-extract with a polar solvent.

In a similar way, polar material that passes through a reverse-phase cartridge will adhere to a hydrated silica cartridge column. Applying your sample directly to an untreated silica SFE and eluting with hexane will wash out very nonpolar compounds. A step gradient with increasing amounts of chloroform, then methanol, and finally water will remove almost everything from the cartridge. A notable exception are proteins that adhere by ion-exchange as well as polar interactions. At neutral pH they seem to bind almost irreversibly. Passing an aqueous protein solution down a silica cartridge and washing with water is an excellent way to remove proteins if you do not exceed the column's capacity.

SFEs are available to use with other chromatography modes to achieve size, ion-exchange, and affinity separations. If the compound to be separated contains a charged functional group or a group with an inducible charge, such as an amine, it can be removed from neutrals and compounds with the opposite charge. Charged molecules can also be removed with ion-exchange resins loose packed above glass wool in a Mohr pipette. Apply the sample to the ion exchanger in a low salt buffer at a pH to maintain the target charge. Wash out the breakthrough, then elute the compound of interest with either high salt or with a pH sufficient to neutralize the charge on the compound or the ion exchanger.

Although organic solvents may aid in making your compound soluble, be careful. Many polymeric ion exchangers will collapse in the presence of more than 20% organic solvents. The bonded phase silica ion exchangers will not tolerate pH below 2 or above 8, but are resistant to organic solvents.

Size separations are seldom run in SFE columns because the column beds are not long enough to give an effective size cut. The only exception to this is in removing salt in a buffer solution from a protein fractionation. Cartridge columns containing Sephadex G-25 can be used as desalting columns.

Affinity SFE are coming on the market for antibody preparation (protein G and protein A columns), for hydrogenase preparations (blue, red, and green dye columns), and activated cartridges that can bind proteins and amino-containing compounds to prepare affinity cartridges are available. These may go beyond the basic definition of the SFE, but they are just simple variations of the same theme.

12.1.4 Extracting Encapsulated Compounds

The final types of materials that may cause problems in extraction are membrane-bound or encapsulated compounds. In the past these have been removed by adding detergent to break the membrane, pulling everything into solution and then extracting out the compound of interest. It is difficult to get clean separations and the chromatography is often contaminated by the detergent.

A better method is to add an equal volume of dimethyl sulfoxide (DMSO) or dimethyl formamide (DMF) to the aqueous sample. This breaks membranes and pulls both polar and nonpolar material into solution. The second step is to dilute the sample with 10 volumes of water. At this point, nonpolar compounds can be removed by solvent extraction or with a C_{18} SFE column. Charged molecules can be recovered with pH-controlled extraction or with ion-pairing reagents. The DMSO or DMF stays with the water layer. Customers have told me they can achieve simultaneous quantitative recovery of both fat-soluble and water-soluble vitamins from encapsulated mixtures. Vitamins are encapsulated to protect potency from air oxidation. Water-soluble vitamins have nonpolar encapsulation; fat-soluble vitamins have a polar encapsulation. It makes them difficult to extract at the same time except through this technique.

It should also offer promise for cell extractions, which, after all, are lipid/protein membrane-encapsulated mixtures of polar and nonpolar compounds. It would be interesting to see the effect of DMSO/DMF on proteins. My guess is that proteins might denature in 50% DMSO and precipitate so they could be filtered off or might renature and refold on $10\times$ dilution in water and stay in solution. This might make a useful research problem.

12.1.5 SFE Trace Enrichment and Windowing

To get the most out of the HPLC as a "time machine" we have to analyze the separation, then put to work the extraction techniques we have discussed to speed the separation. As an example we will take a metabolite study carried out by one of my customers. His problem was to study the dispersion of a compound, XX, in the environment. This might involve looking at samples of agricultural runoff water, river water, sludge, soil, and fish tissue.

In examining environmental runoff water he ran into a problem. His samples were very dilute, so the first step would be some form of concentration. Studies with the pure material show that it was sufficiently nonpolar to stick to a C_{18} cartridge from an aqueous solution.

Step 1: Concentration and Gradient Elution

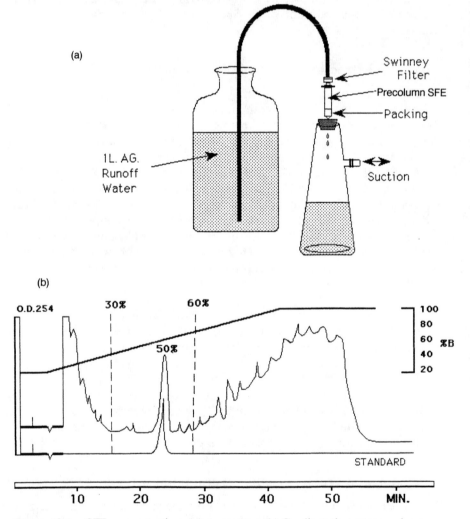

Figure 12.1. SFE concentration. (a) Apparatus; (b) Gradient chromatography.

He prepared an SFE cartridge by first wetting it with 2 ml of methanol and then with 2 ml of water. Next, the SFE was attached below a particulate filter in a line leading to a 1-liter suction flask (Fig. 12.1a). The other end of the filter was connected to tubing dipped into a 1-liter flask of the runoff water containing our compound, XX, at 5 parts per trillion. (He was unable to detect this level of compound by direct injection into his HPLC system.)

The runoff water was pulled through the SFE cartridge by suction leaving the sample adhering to the bonded-phase support. He used a 2-ml water wash to remove adhering polar material and the sample was eluted with 2 ml of acetonitrile. This is a 500-fold concentration increase in going from 1 L to 2 ml. He evaporated the sample, dissolving the sample in starting mobile phase with sonication, injected the sample, and made a separation with a 45-min analytical gradient HPLC run obtaining a peak corresponding to the standard, XX, with the correct peak height (Fig. 12.1b).

To check recovery of compound from the medium, he ran a labeling study. Radiolabeled sample was added to runoff water at 5 ppt, it was sonicated, extracted with a wetted SFE, eluted as before, and counted. He found a 97% recovery of radiolabel from the SFE.

The problem with the separation so far is that he had increased concentration, but not improved his run time. Even though he was interested in only a single component, he had to run a gradient to separate it from the rest of the mixture. He still must prevent early running, more polar compounds from running into the sample and washout late runners before the next injection. His bottom limit on run time was about 45 min with a 15 min equilibration to achieve reproducible results.

An hour per run is a long time if you have many samples, as would be expected in a metabolite study. His next step was to simplify the chromatograph conditions by improving the sample preparation step.

At this point I came into the picture. Together we repeated the loading steps to the point where the sample was on the C_{18} SFE (Fig. 12.1a). Going back to the gradient run (Fig. 12.1b), we see that the sample peak came off at 50% acetonitrile in the gradient profile. We decided to change the recovery step of the SFE cartridge. First we washed the cartridge with 2 ml of 30% acetonitrile (Fig. 12.2a). Next, we eluted with an equal volume of 60% acetonitrile and finally with 100% acetonitrile into fresh test tubes.

To check the recovered fractions, we equilibrated the HPLC with 60% acetonitrile. The standard, XX, and the first cartridge eluant (30% cut) were run under these conditions (Fig. 12.2b). The first cut chromatogram was examined for the presence of a peak with the same retention time as the standard. The remaining SFE elution fractions (60 and 100% cuts) were run at the same conditions and examined for peaks eluting with the standards retention times (Fig. 12.2c and 12.2d).

We were trying to determine the location of the sample peak in the eluant cuts. Ideally, it should all fall in the "window" represented by the middle or second cut (Fig. 12.2c). If we find standard in the 30% eluant, then we have to

Step 2: Windowing

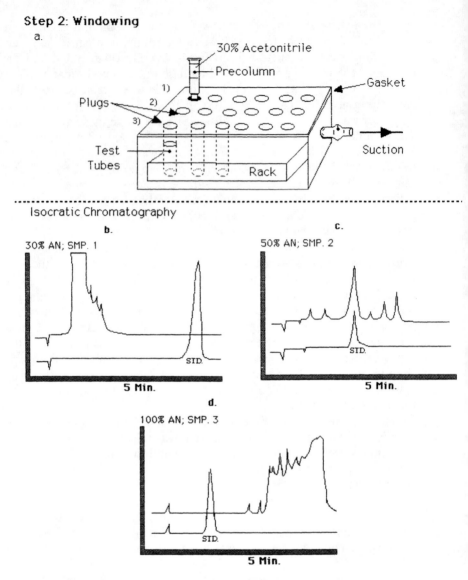

Figure 12.2. SFE windowing: (a) Apparatus; (b)–(d) Isocratic chromatography; (b) fraction 1; (c) fraction 2; (d) fraction 3; (AN = acetonitrile).

move the window frame to exclude the sample from the 30% cut. We do this by repeating the concentration step, then using 25% instead of 30% acetonitrile for the first wash. If the XX had overlapped into the 100% eluant fraction (Fig. 12.2d), a similar window frame movement could have been made by increasing the 60% (cut 2) fraction to 65%.

Once the window was optimized for the compound of interest, only the "window" eluant need be run for each sample. Instead of a gradient, we could use a fast running isocratic. We eventually got the separation down to a 5-min run time. The 30% "wash" and the 100% "stripping" eluants were run only to optimized the "window" and to prove reproducible recovery. They were discarded for later runs.

To determine the effectiveness of the windowing procedure, the customer went back to his radiolabeled standard. Radioactive recovery studies showed better than 99% release from the cartridge and 94% activity recovery in the main "window" for the compound under study.

Using these techniques we were able to increase the concentration by $500\times$, while decreasing run time from the gradient to the isocratic chromatography by $12\times$. This is the type of simplification that can allow the full power of an HPLC to come into play.

12.1.6 Derivatives

Derivatives are used in HPLC only as a last resort. HPLC derives its usefulness from being able to run aqueous samples directly, with little if any work-up. One place where derivatives are necessary is in preparing detectable forms of compounds that have poor absorption by themselves. Carbohydrate, fatty acids, lipids, and amino acids all have poor UV absorption. Useful derivatives for fatty acids and amino acids already exist; others should appear in the near future. Fatty acid derivatives are made with bromphenacyl bromide in strong base; the resulting mixed anhydrides can be detected at 50 ppt in a good UV detector. Amino acids can be removed from peptides in the Edmund degradation reaction as phenylthiohydantoin derivatives (PTH amino acids). Aqueous amino acids can be derivatized with o-phthaldehyde (OPA) and be detected with a fluorescence detector in the presence of excess reagent. Much work has been done on postcolumn, in-line OPA amino acid derivatization with a reactor between the column and the detector, but none works very successfully.

12.2 Methods Development

There are a number of methods for selection of conditions when approaching a new separation. You can rely on published methods in the literature (sometimes an extremely dangerous step), or you can use published methods as a guide to conditions. If the compound has not been published, you can rely on methods published for compounds of similar polarity. You can even estimate conditions from consideration of the compound's structure.

All of these tend to be rather hit or miss. For unknown compounds or unusual mixtures, they are generally not very successful. It is usually better to approach a separation from a more systematic methods development based on scouting gradients (see Chapter 3).

When I demonstrate an HPLC system, it is often necessary to develop a separation of an unknown mixture of compounds in half a day to aid in obtaining an instrument sale. The following techniques arose from the need to speed the achievement of such separations.

12.2.1 Standards Development

We are now going to walk through a typical methods development as you might carry it out in a clinical laboratory. Let us assume that you have just started your first day on the job. Your brand new gradient HPLC is sitting on the laboratory bench with installation just finished. The Chief Pathologist for the hospital, your boss, walks in and hands you four vials marked A, B, C, and D. You are told that you will be receiving patient blood samples containing any or all of these four compounds at 10:00 A.M. tomorrow. Your boss wants the results tomorrow before leaving the hospital.

What are you going to do? Where will you start? You need to develop a rapid analysis for four compounds from plasma. The first step is to prepare $100\times$ standard solutions for making dilutions. You make a carefully weighed 1-g/L solution of each compound in acetonitrile. A quick UV scan tells you that all compounds absorb well at 254 nm. Part of each is stored in the freezer; the rest is retained for development work.

To make an injection solution, you add 1 ml of each $100\times$ to a 100-ml volumetric flask and dilute with 25% acetonitrile/water, the starting mobile phase. You start the gradient system equilibrating a guard column connected to a 25-cm C_{18} column at 2 ml/min in the same mobile phase. You inject a 7-μl sample and run a 20-min gradient. Your first peak comes off at 60% on the gradient trace. You get three partially resolved peaks (Fig. 12.3a).

Next, you drop back 10% from where the first peak came off and dial in 50% acetonitrile (AN) water. You let the column equilibrate for 10 min and shoot the next sample. You get three well-resolved peaks, the last of which tails badly (Fig. 12.3b). There is no indication of a shoulder indicating an unresolved peak. Either one compound is very nonpolar or it is coeluting with one of the other compounds. You need to make an α change.

You change the stronger solvent by switching to MeOH, while maintaining the same polarity. Since you used 50% AN/water, you now switch to 65% MeOH/water, equilibrate the column, and inject a new sample. You now have four resolved peaks, with the third tailing into the fourth (Fig. 12.3c). Guessing that the ionization problem is due to an acid, you add 1% acetic acid to the mobile phase and inject again (Fig. 12.3d). You get lucky and have four sharp, resolved peaks. They are a little farther apart than needed for the integrator, so you switch to 70% MeOH/water with 1% acetic acid to draw them together and bring them off faster.

Note 1: Acids are easier to deal with in scouting so you lowered the pH with 1% acetic acid.

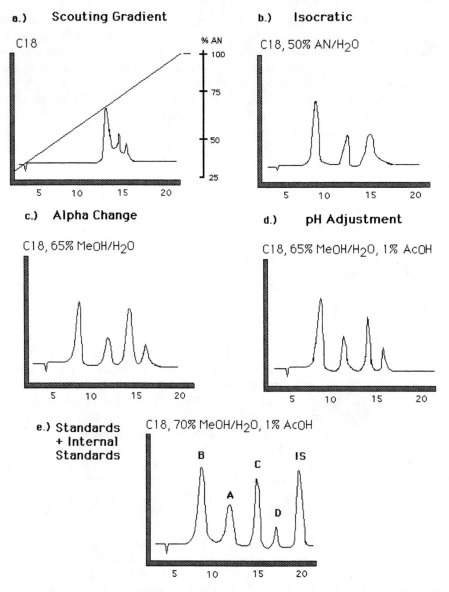

Figure 12.3. Systematic methods development: standards. (a) Scouting gradient; (b) Isocratic; (c) Alpha charge; (d) pH adjustment; (e) Standards plus internal standard.

Note 2: Amine anions are more likely to be found in pharmaceutical compounds than acids. If the acetic acid had not worked, you would add nonyl amine to 5 mM to overcome interaction with silica surface cations.

The next step is to identify each peak. You dilute and inject each compound separately and compare retention times to the standard mixture run. The elution order is B, A, C, and D. The last thing you need to find is an internal standard with a retention time just longer than D. You find compound IS and make a 1-g/L solution. You will add 1 ml to each standard mixture before dilution. You run a standard run with IS and calculate response factors for each peak relative to the peak height of IS (Fig. 12.3e).

At this point, if you knew the identity of each component, you could rush out and publish the separation. Many publications 2–3 years ago stopped here. However, the work is only about a third done. So far we have separated pure compounds under ideal conditions; we have still not entered the real world.

12.2.2 Samples Development

To finish the method, we need to add sample preparation and examine our standard separation under real conditions. Since you will be working on blood samples, you need to start looking at the standards in blood. You need fresh blood, uncontaminated with drugs or interfering compounds, and representative of a wide population. The blood bank is out; they stabilize their blood for storage. You obtain volunteers, telling them how they can help push back the frontiers of ignorance and help society. Then you send them to an intern you know.

Once you get your blood samples back, you pool them, then split them into two parts. You refrigerate one part for later work with standards. You mix the other part with an equal volume of acetonitrile, homogenize, sonicate, place stoppered in a boiling water bath for 5 min, and then sediment proteins and particulates by spinning in a centrifuge. You place 2 ml of the supernatant in a volumetric flask and dilute with mobile phase. Running an 8-μl sample gives you a chromatogram of typical denatured blood plasma sample (Fig. 12.4a). You see a broad, early polar peak, which may overrun your first two standards, and a large lipid peak. The column may have to be washed with DMSO/MeOH to get the last of the nonpolar compounds off. The guard column will be collecting most of the albumin fraction proteins and lipids during our runs. It must be periodically disconnected, inverted, and washed into a beaker with strong solvent.

The next thing you want to see is the plasma blank sample with your standards and internal standard added to check for possible interferences. You add 2 ml of your treated blank and 1 ml each of the 100$\times$ solutions of A, B, C, D, and IS to a 100-ml volumetric flask, dilute, and inject. Sure enough, peak B is buried and A is on the shoulder (Fig. 12.4b). You need to clean up the plasma blank with extraction or windowing techniques.

You decide to do windowing on a C_{18} SFE cartridge column. You take the mixture of the four original standards, dilute it 5-fold with water to increase the compound's k's, and pass it through the MeOH and water-wetted SFE cartridge. You elute the cartridge with 2 ml each of 60, 80, and 100% MeOH, add 1 ml of IS to each fraction, dilute 100$\times$, and shoot each sample into the HPLC

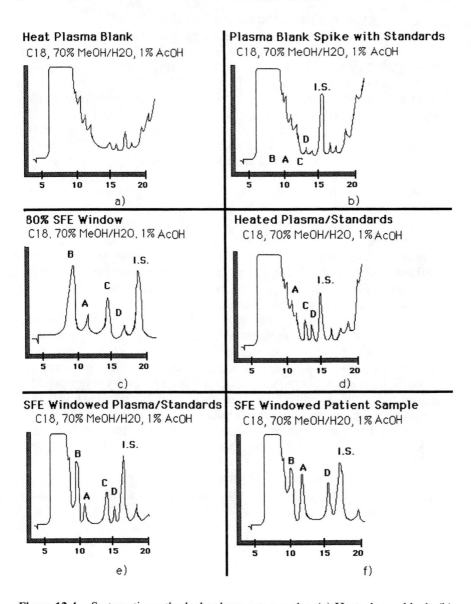

Figure 12.4. Systematic methods development: samples. (a) Heat plasma blank; (b) Plasma blank spike with standards; (c) 80% SFE window; (d) Heated plasma/standards; (e) SFE windowed/plasma standards; (f) SFE windowed patient sample.

system. The 60% wash is contaminated with B and a trace of A. You repeat moving the left window frame to the left by eluting with 55, 80, and 100% MeOH. All your peaks are in the 80% window (Fig. 12.4c); none shows up in either the 55 or 100% washes. Quantification against the internal standard's peak height shows no loss of peaks B or D on the SFE.

You remove the other half of the pooled blood sample from the refrigerator. You mix 4 ml of blood with 1 ml each of the four standards in acetonitrile, sonicate, heat in boiling water, and centrifuge. You place 2 ml of the supernatant and 1 ml of IS 100× in a volumetric, dilute with mobile phase, and inject into the HPLC. (*Note:* you are looking for loss of standards by adherence to precipitated protein.) Peaks A, C, and D are present; the last two quantitate correctly (Fig. 12.4d). You take 2 ml of the remaining plasma plus standards supernatant, dilute it 5-fold with water, and place it on an activated SFE cartridge column. You then elute with 55, 80, and 100% MeOH in water containing 1% acetic acid. The 80% fraction is mixed with IS, diluted, and run. It shows a much narrower plasma polar peak, compound B as a shoulder on the plasma peak, peaks A, C, D, and IS, and a small amount of late running nonpolar peaks (Fig. 12.4e). All four standards give correct peak height response factor to the IS peak. You are ready to accept patient samples.

Two more comments are necessary. The internal standard, IS, is usually added to correct for injection variations. The way it was used in the last step, it was also checking for standards recovery from the protein precipitation step. It is mildly dangerous to use the same internal standard for two purposes. If the quantification was not correct, it would have been necessary to repeat both the injections and the precipitation with another internal standard to find the problem. Also, we must check for possible interfering drugs (ones coeluting with our standards) that might be given to patients taking these drugs. I would use the plasma blank spiked with standards and IS to look for these interferences by changes in response factors of the standards.

To run a patient sample, you go through exactly the same deproteination, SFE cartridge extraction, IS addition, mobile phase dilution, and injection steps (Fig. 12.4f). From the peak heights relative to the IS height, we can now quantitate the amount of each drug in the patient's blood. To ensure linearity, you need to dilute the windowed plasma blank and spike it with different levels of each standard and plot calibration curves for each compound, but, basically, our methods development is done.

12.3 Gradient Development

It is sometimes not possible to develop an isocratic separation for complex mixtures of compounds. Binary methods development starts with a linear gradient from 25 to 100% acetonitrile over 20 min (Fig. 12.5a) just as we did in scouting for an isocratic standards development. Next, inspect the gradient for areas in which peaks are jammed together (h-1, h-3) and areas in which they

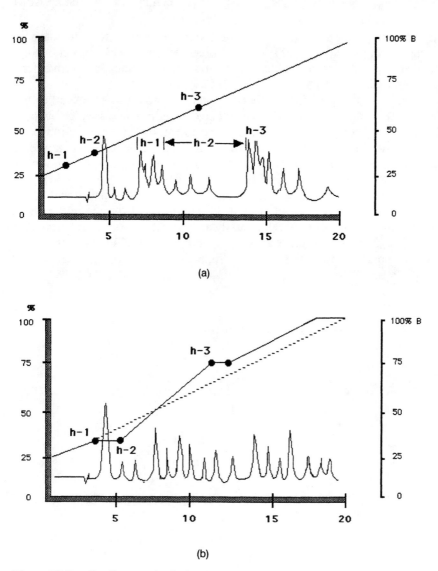

Figure 12.5. Gradient methods development. (a) [Title TK]; (b) [Title TK].

spread too far apart (h-2). Imagine that there is a *hinge point* 10% before each of these points in the gradient trace.

If peaks are pushed too close together or unresolved (h-1, h-3), place a hold in the gradient equal in length to the time in the original gradient for the last compressed peak to come off, then return to the original slope. If peaks are too far apart, go back to the 10% hinge point (h-2) and increase the slope of the gradient so the last peak will be reached in half the time, then return to the original gradient slope (Fig. 12.5b). Good scientific procedure would have you change one point at a time. I have been successful, however, in changing a number of points, rerunning the chromatogram, checking for improvements, and then making new changes.

These are simply guidelines that I have found to be useful. The key is to decrease the gradient slope before compacted areas and increase it before peaks spread too far. I use the scouting gradient to improve the separation before going to the slower, more reproducible analytical gradient. You can also use acetic acid or nonyl amine to sharpen tailing peaks. Add equal amounts of either additive to both the A and B gradient solvents.

If you already have an analytical gradient, but are not satisfied with it, you can still use hinge point development. Figure the dead volume in the system from pump to detector flow cell (usually around 1.5 ml for pump, mixer, and injector, 2–3 ml for column void volume and detector). Subtract this volume from the point you want to change and make this your hinge point.

13

Application Logic: Separations Overview

At this point, I am going to try something a little different. Most HPLC texts include a series of figures showing separations, including the conditions, for various classes of compounds. I prefer to give you tools to predict new separations. First, to give you an approximate set of conditions for making almost any type of separation, and second, to indicate why a particular column, mobile phase, and detector (or wavelength) was chosen for this separation, *the logics of the separation.*

To address the first objective, I have included a separation guide (Appendix A) designed as a quick reference to conditions that could be adapted to separate compounds similar in polarity, in size, in charge, or in absorption. Where possible, isocratic runs were chosen, rather than gradients. To handle the second objective, we will discuss the various classes of materials exploring the chemical and physical differences that dictate certain HPLC conditions.

13.1 Fat-Soluble Vitamins, Steroids, and Lipids

The first grouping is a mix of fat-soluble compounds that function as hormones, cofactors, and membrane components. *Fat-soluble vitamins* separate on a C_{18} column in 80% acetonitrile/water and are usually detected at 280 nm or with fluorescence. *Triglycerides* are slightly less nonpolar than fat-soluble vitamins and require 60% acetonitrile/water to run on C_{18}. They have poor extinction coefficients, and detection at 220 nm only competes with refractive index detection in sensitivity. A phenyl column run in 50% acetonitrile/water gives some separation based on double-bonded side chains. *Steroid hormones* require similar conditions and separate on a C_{18} column in 60% MeOH/water. At 230 nm, estrogenic steroids can be detected at 150 ng (the level in a pregnant woman's urine). Adrenocorticosteroids have higher extinction coefficients and can be seen at 240 nm. *Prostaglandins* are hormonal, aliphatic diacids with double bonds in their structure. They are separated on C_{18} in 35% acetonitrile (AN)/water containing phosphoric acid at pH 2.5. The phosphate is needed as a buffer, since detection is at 192 nm, almost the bottom limit of the UV detector in air; even then not all of the prostaglandins absorb. These isocratic conditions will separate most of the common prostaglandins, but you will have to use a gradient up to 100% acetonitrile (AN) to separate all of them, up to and including, arachadonic acid, the precursor for the prostaglandins.

The final type of "fatty" compounds in this group, *phospholipids,* are the hardest to separate. They are naturally occurring "soaps" with long side chains containing alcohol, sugar, or sugar alcohol bodies, and charged phosphate groups. They are soluble in nonpolar solvents when extracted from acidified media, but they differ in polar functional groups. They have very poor UV absorption and must be detected with "end absorption" at 206 nm. The most successful separation has been on an acidified silica column with isocratic elution in 4% MeOH/AN containing 1% phosphoric or sulfuric acid.

13.2 Water-Soluble Vitamins, Carbohydrates, and Acids

Water-soluble vitamins have a range of polarities. The B-complex, except for B_{12}, can be separated on a C_{18} column in 8% AN/water at 280 nm using heptane sulfonate as an ion pairing reagent. The ion pair slows thiamine and nicotinic acid so they will retain and run close to riboflavin. Vitamin C, an oxidizable, organic acid, separates on C_{18} with 5% MeOH/water adjusted to pH 2.5, but has poor UV absorption and is better detected electrochemically for high sensitivity. All vitamins, except C and B_{12}, can be seen at UV (254 nm). B_{12} also has poor UV absorption, and may be a good candidate for high sensitivity conductivity detection when it is available. It has a central cobalt atom that might be detectable at the right voltage with an electrochemical detector.

Free fatty acids can be separated on a C_{18} column based on carbon number using 50% MeOH/water at pH 2.5. A fatty acid column (actually a phenylalkyl column) will also separate them based on the number of double bonds. Fatty acids can be detected at 210 nm with a UV detector or by using a refractive index detector. For high sensitivity work they are derivatized with bromophenacyl bromide and separated on C_{18} in a 15–80% AN/water gradient at 254 nm. Increases in early running C_2 and C_4 fatty acids are used as indicators of bacterial action. *Krebs cycle acids* are di- and tricarboxylic acids involved in metabolism of fats, sugars, and amino acids. They are separated by anionic ion exchange on an amino column using a pH 2.5 buffer gradient from 25 to 250 mM phosphate with detection by refractive index detector. If sensitivity is required they could be derivatized with bromphenacyl bromide.

Monosaccharides can be separated on a polymeric cation-exchange column with a pair-bonded calcium or lead ion. The mobile phase is 80°C water and detection is either by refractive index detector or by UV at 195 nm. The elevated temperature speeds equilibration in the polymer column and reduces viscosity. Detection sensitivity is poor and numerous attempts had been made to prepare high sensitivity derivatives. This column can separate positional isomers, such as glucose and galactose, ring isomers, such as glucose and fructose, and all of these from polysaccharides and sugar alcohols.

Polysaccharides can be separated also on amino columns run in 75% AN/water and in polymeric "carbohydrate" size separation columns in 80°C water with detection at 195 nm. The amino column separation can only go to about decasaccharides with 10 sugar groups. The size separators can go to MW of about 6 million and offer separations of large polysaccharide that have been separated previously only by crystallization. A small amount of organic solvent will sharpen separations on either of the polymeric carbohydrate columns, but must be kept below 20% concentration to avoid damage to the column packing through swelling or shrinkage. Heating the water reduces column backpressure by decreasing viscosity.

13.3 DNA Family: Nucleic Acids, Nucleosides/Nucleotides, and RNA/DNA

The *nucleic acids* are a family of compounds ranging from simple purine and pyrimidine bases to sugar and phosphate containing nucleosides, nucleotides, and polynucleotides, such as RNA and DNA. The nucleic acid ring structures all absorb well at 254 nm in a UV detector. The free *nucleic acids* have been separated on a cation-exchange column using high levels of ammonium acetate at pH 4.6. Most show pK_as at 3–5 and might give sharpened peaks at pH 2.5.

Nucleosides, which have sugars connected to the nucleic acid ring, have been separated on a C_{18} column in 8% MeOH/water at pH 5.5 with phosphate. Adding the phosphate groups to form mono-, di-, and triphosphate *nucleotides* increases solubility and they are separated with a quaternary amine ion pairing reagent, tetrabutylammonium phosphate. A C_{18} column is run in 20% An/water pH 2.65 containing 10 mM TBA. Phosphate concentration is controlled at 30 mM; greater than this leads to loss of di- and triphosphate resolution.

Polynucleotides pose a separation problem because of their large sizes and long, rigid shapes. tRNAs and some bacterial DNAs, which form ring structures, can be separated on large pore, TSK-type, size-separation columns. Mammalian mRNAs and DNSs are double helix molecules that form rigid rods with large Stokes radii. A size column with a 2 million MW exclusion for proteins will exclude DNA restriction fragments larger than 100,000 MW.

Added to this is the fact that nucleic acids are fragile and that pressure shearing on silica packings has been reported. Genetic engineering research has given this area considerable importance, and new solid-bead, ion-exchange columns are just now emerging for separating larger nucleic acid sections. Purification of cloned restriction fragments and removing contamination from DNA amplifications reactors are increasingly important applications for HPLC systems.

13.4 Protein Family: Amino Acids, Peptides, and Proteins

Separation of the family of compounds leading to enzymes and structural proteins has been an area of much recent research. *Amino acids* show "end absorption" at 220 nm, but not high extinction coefficients. If a particular amino acid has a chromophore in its side chain it may absorb well at higher wavelengths; Phe and Tyr absorb strongly at 254 nm and Trp at 280 nm. The peptide bond between adjacent amino acids has good absorption at 210 nm in peptides and proteins.

Amino acids are derivatized two ways to increase sensitivity. *Free amino acids* in solution are reacted with *o*-phthaldehyde (OPA) to form a fluorescent derivative that excites at 230 nm and emits at 418 nm. These OPA derivatives are separated on a C_{18} column in a complex mixture of AN/MeOH/DMSO/water pH 2.65. *PTH amino acids* are formed from the N-terminal end of peptides during Edman degradation for structure analysis of proteins. HPLC is used to identify which amino acids are released. PTH amino acids are separated at 254 nm on a C_{18} column with a gradient from 10% THF/water with 5 mm AcOH to 10% THF/AN. The separation with reequilibration takes 60 min. Work with short 3-μm columns has reduced this separation to 10-min gradients.

Peptides (<99 amino acids) are separated at 254 nm on C_8 in 30% *n*-BuOH/water containing 0.1% trifluoroacetic acid (TFA). They can also be separated in acetonitrile/water gradients in which 0.1% TFA is added to both water and AN. Peptides up to 30 amino acids can be separated on C_{18} and up to 99 amino acids on C_8. (Avoid going over 70% AN in the gradient. TFA is reported to form aggregates in acetonitrile concentrations greater than 70% resulting in very large baseline shifts.) Peptides can also be separated at 210 nm on a C_3 column using AN/water gradients buffered with phosphate pH 5.5; these conditions are especially important if the peptides do not contain aromatic amino acids.

Enzyme *proteins* are separated with retention of activity in most cases on a TSK-2000sw size-separation column in 100 mM Tris pH 7.2 with 100 mM NaCl with detection at 280 nm. Phosphate and sulfate will also work, but peaks are sharper with chloride. Protein stabilizers such as glycerol, EDTA, and dithioerythritol can be added to the mobile phase. Enzymes can also be separated at pH 7.5 on TSK DEAE and CM ion-exchange columns using salt gradients to 150 mM NaCl. DEAE is usually the first choice over carboxymethyl.

Antibodies and larger proteins can be separated on TSK-3000sw columns. Proteins for structural studies can also be separated under denaturing, partition condition. A C_3 column can be used in 0.1% TFA gradient to 70% AN/0.1% TFA. Proteins with large nonpolar groups, such as albumins, tend to stick very tightly to this last column. Resolution increases in the order size $<$ ion exchange $\ll$ partition. Load increase in the order partition $<$ ion exchange $\ll$ size.

13.5 Clinical Drug Monitoring, Toxicology, and Forensics

Drug monitoring tends to be of two types: (1) assays for specific drugs, closely related analogs, and preparation enhancers and (2) rapid, broad screening for overdosage detection of drugs of abuse.

Theophylline, an asthma controller, has a very low safety/therapeutic ratio. One of the first clinical application for HPLC was to titrate theophylline levels in patient plasma to avoid toxic overdoses. Blood levels can be controlled by assay at 270 nm on C_{18} in 7% AN/water at pH 4.0 with phosphate.

Catecholamines, nerve transmitters monitored in brain and heart patients, are separated on C_{18} using octane sulfonate ion pairing in 6% AN/water pH 3 with EDTA and phosphate. Detection can be at UV (270 nm) or by electrochemical detection at $+0.72$ V for maximum sensitivity. Other tyrosine and tryptophan metabolite neurotransmitters such as serotonin, VMA, and HMA can be analyzed with ion pairing and EC detection.

Anticonvulsants, used in controlling seizures, are analyzed on C_{18} at 220 nm eluting with 40% MeOH/water. They are also common drugs of abuse and are monitored for in toxicology laboratories.

Tricyclic antidepressants, major tranquillizers used in hospitals, are separated at 254 nm on C_{18} using 55% An/water pH 5.5 with pentane sulfonate. Since these are very basic compounds, it is necessary to used endcapped columns. Their separation benefits from organic modifiers, such as nonyl amine.

Basic drugs of abuse can be screened in a toxicology laboratory using a 20-min gradient from pH 3.0 phosphate buffer to 25% AN/buffer at 215 nm. Similar screens can be set up for acidic drugs such as the barbiturates. Identity confirmation is very important in these laboratories to avoid false positives. This can usually be done using a scanning or wavelength ratioing detector or using multiple chromatography detectors in series.

13.6 Environmental and Reaction Monitoring

HPLC serves for some monitoring of air and water pollution, with much more planned in the near future. Air quality can be determined by pulling known volumes of air into an evacuated metal chamber and analyzing with a GC or into a prewetted C_{18} SFE cartridge column, then eluting under windowing conditions and analyzing on the HPLC. This has been used with belt monitors or stack gas pumps to analyze laboratory exposure in toxic or radioactive environments. Water pollution can be monitored in the same way. Instead of storing gallon bottles of water, the water can be pumped through a cartridge column, placed in a plastic bag, and refrigerated or frozen for later assay.

Pesticides and polynuclear aromatics (PNAs) are the most commonly analyzed environmental contaminants. PCBs, dioxans, and nitro organics (explosives) are of growing importance. The major obstacles to adoption of environmental HPLC application are (1) awareness of the need (i.e., environmental runoff and drinking water contamination), and (2) the slow rate of development and acceptance of new AOAC and EPA HPLC-based methods.

Pesticides can all be analyzed on C_{18}, the chlorinated hydrocarbon type (Chlordane) at 80% AN/water (220 nm), the carbamate type (Sevin) at 40% AN/water (254 nm), and the organic phosphate (Malathion) at 50% (192 nm). The organic phosphate types are hard to detect at low concentration and various postcolumn phosphate analysis techniques have been evaluated. LC/MS offer considerable promise for analyzing all of the pesticides in a general gradient HPLC scheme.

PNAs are analyzed at 254 nm on C_{18} in 80% AN/water. PCB can be analyzed with the same conditions. Dioxans require detection at 220 nm and separation in 50% AN/water.

Reaction monitoring is an important use for HPLC, both in the laboratory and in production. I have followed reactions in a 500-ml stirred round bottom flask and in a stirred 6000-gallon reactor. The HPLC becomes your window-on-the-reactor. It lets you analyze starting materials (QA, quality assurance), your basis for yield, reaction intermediates, by-products, and final product (QC, quality control).

In one study, I followed the temperature breakdown of the final product by stirring it in while stripping solvent with live steam. I used an inverted pipette to pull a sample, weighed it in a weighed flask, threw in chloroform, shook it, pulled a sample of the chloroform layer, spun it to break the emulsion, then shot it into the HPLC. I ran 5 min behind the reaction with my "reaction window." I could follow the disappearance of starting material versus heating time. Using other sampling techniques, we could follow product from the reactor, into holding tanks, through steam strippers, through dryers, into product bags. It is simply a matter of scale; the analytical tool is the same.

13.7 Application Trends

What can we get out of all this that will help us the next time we run into a new separation?

First, *mobile phase and column:* In Appendix A we will see that most of the small molecule separations could be made on C_{18} in acetonitrile/buffered water, with the exception of charged molecules and carbohydrates, which are too water soluble. We will see a range of polarity from fat-soluble vitamins, steroids, triglycerides, and chlorinated pesticides eluting in 60–80% An/water, to carbamate, phosphate pesticide, anticonvulsants, and antidepressants at 40–50% AN/water, to nucleosides, nucleotides, aspirin, and water-soluble vitamins at 5–10% AN/water. If you know something about the compound's structure or its solubility, you have a good clue as to what mobile phase can be used for its separation.

Second, *detection:* Colored compounds absorb above 300 nm, and compounds with aromatic groups will probably absorb around 254–280 nm; if it has a carbonyl group it should absorb at 220 nm. Any solvent or buffer with an oxygen in it is almost useless below 220 nm unless you can run isocratic and zero out the solvent absorption. Water, acetonitrile, and phosphate are the choice for UV detection at very low wavelengths (<200 nm). Refractive index detection finds most compounds, but not at high sensitivity; it is good for preparative work, but cannot be used for gradients. Fluorescence detectors give you high sensitivity, *if* your compound fluoresces and if you can find the wavelength combination for excitation and emission. Electrochemical detectors give excellent sensitivity, but work only with oxidizable or reducible compounds and can be used only with <25% organics because of flow cell and reference electrode design.

It does not sound like much help, but with these as guides, and a little intelligent thought, you should be able to separate most of the compounds you encounter. The rest are research projects and, therefore, publishable, so they are not a complete loss.

14

Automation: Separations and Data Acquisition

14.1 Analog-to-Digital Interfacing

In the real world, changes are continuous *analog* variations with time. Temperatures rise and fall. Instrument parameters, such as signal output voltage, increase and decrease continuously. In the computer world changes are discrete *digital* steps; computers process using a two value alphabet, 0 and 1. The next value may or may not be related to the one before it.

There are good examples of both analog and digital devices in electrical light switches. The light switch you use to turn the room lights on and off is a digital device; it has two discrete settings. The dimmer switch you use to provide a romantic setting for a dinner party is an analog device; it has a continuously changing output.

For analog instruments to communicate with computers the analog signal must first be converted to discrete digital steps (digitized). This is the purpose of the analog-to-digital (A/D) converter (Fig. 14.1.a). Most analog instruments put out a variable voltage signal. Strip chart recorders work on an analog voltage that varies from 0 to 10 mV. Baseline is usually adjusted to 0 V and 100% full scale to 10 mV. As a detector measures a change in the flow cell due to a peak, the detector electronics outputs an equivalent signal change that increases then decreases in voltage. The strip chart traces this signal on paper to produce the peaks of a chromatogram.

An integrator or a computer usually requires a stronger analog signal from the detector, either 0–1 or 0–10 V, but the output is of the same form. As the signal reaches the A/D board it is sampled for a specific length of time and the

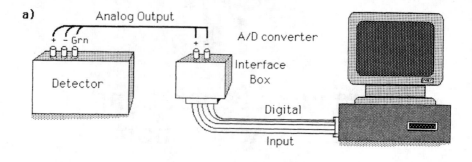

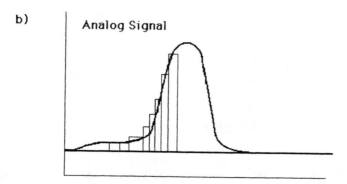

Figure 14.1.   A/D conversion. (a) Analog output; (b) Analog signal.

strength of the signal is converted to a digital number. This is done on some boards by charging up a capacitor, shorting it, and then charging it again. The number of times it charges in a specific time period is represented by a digital number. The slower it charges, the smaller the number; the faster it charges, the larger the number. These digital values are represented by the stair steps in Figure 14.1.b.

The frequency of this sampling is defined in samples per second. Most A/D boards can sample at about 30,000 points/sec, while an HPLC signal requires sampling at no more than 10 points/sec. The other controlling variable is the size of the output digital number. An 8-bit A/D board can only process numbers up to 65,000. To process the full range of detector outputs, at least a 12-bit board is required and many boards use a 16-bit data path that allows a word size that can handle numbers up to 1 million.

Once the signal is converted to a digital number, it can be transmitted to the computer along with a number representing the sampling rate. Our signal now consists of a sample rate number followed by a series of digital numbers representing the voltage strength at each subsequent time point up to the end of sampling time.

This would be equivalent to taking a motion picture and breaking it into a series of individual film frames. Each frame represents a digitation of the motion picture at a given time point. Running the frames through the projector fast enough gives the appearance of a continuous, analog presentation. The digital-to-analog (D/A) converter produces a similar effect with a digitized signal. It takes the sampling rate and the series of digital numbers and converts them into an approximation of an analog signal. Smoothing processes produce a continuous signal. This signal can then be sent to control a device that requires a continuously varying voltage signal. Using such a D/A converter, an HPLC controller can send a digital signal to change the motor speed on an analog HPLC pump.

14.2 Digital Information Exchange

The A/D board can reside either in the detector or in the computer. If the board is in the computer, it will have analog input terminals similar to a strip chart recorder or it will be connected to an interface box that will have this type of connector. If the A/D card is in the detector, the detector will have some type of digital interface port and cable to connect it to the computer. In the personal computer world once the signal is digitized outside the computer, it will be sent to the computer over one of three types of computer communication cables: parallel, serial, or GPIB.

Parallel cables transmit each bit of the digital word on a separate line within the cable. Serial cables code the digital words so they can be sent over a single line within the cable. Other lines are used in both parallel and serial cables for communication housekeeping. Both cables are designed to send information only one direction at a given time or asynchronous communication. Parallel cables transmit data faster than serial cables, but are limited to less than 20 ft in length. Serial cables can carry signal up to 1000 ft, and longer with signal boosting.

The GPIB or IEE-488 cable is a bidirectional serial cable carrying module addressing information lines as well as housekeeping lines. It can be used to send and collect information from different instruments on a cable bus. It is often used in instrumentation to automate a series of discrete instrument modules through a system controller.

14.3 HPLC System Control and Automation

HPLC systems operate on a master–slave arrangement in setting up for automation. One module sets the timing and initiates processing, and the remaining modules accept a signal and follow the leader. In a gradient HPLC system, the master module can be a microprocessor-based system controller, a computer software-based controller, an autosampler, or an integrator.

A system controller sends out pump speed control signals to set flow rates and to generate solvent gradients. It should also be able to send 0- to 5-mV contact closure signals to actuated external switches and equipment. It also may be able to send out a 0- to 30-V powered contact closure signal needed to run solenoid valves used to control air pressure and liquid flows. Finally, the controller should be able to take in external 0- to 5-mV contact closure switching signals to control its internal functions (i.e., start a run, stop, pause, etc.). A smart controller may also have a serial or GPIB interface port to allow upload/download of method control programs.

Other smart system modules that may be able to serve as system master are the autosampler or the integrator. In an isocratic system with no pump controller, an integrator might take a contact closure signal from a manual injector as a signal to start data collection and processing. A more complex system might be constructed using a master autosampler to initiate a series of washes, injections, and multiple injections from the same vial then send start/stop signals after every injection to the integrator to start data collection and processing. The integrator in turn can send back a hold signal to prevent further injections until it finishes printing out results. The autosampler might also send vial identification information to the integrator to be included in the printout. A third automation scenario might involve a master autosampler, a gradient controller, a smart detector, and an integrator interfaced to a data storage computer. Complete automation would have the autosampler making an injection and sending a signal to the system controller to start its gradient run and to the integrator to begin data collection, the system controller could send signals to the smart detector to change wavelengths or sensitivity during the gradient run, and the integrator could accept a vial I.D. and transmit it along with its integration output to the computer for data storage and further processing in other software modules to generate a final run report.

In general, in a system with an autosampler, this module initiates run start. The system controller usually controls external events occurring during a gradient run. The integrator may send a hold further injection signal until it finishes printing its processing report.

14.4 Data Collection and Interpretation

To quantitate data in either a computer or an integrator, you must first establish a baseline, then acquire data from an injection, detect peaks, integrate the peaks, and compare the peak integrations to known amounts of standards compounds.

14.4.1 Preinjection Baseline Setting

Before making an injection for a chromatogram that will be integrated, it is necessary to define an integration baseline. This can be done automatically in many integrators by pushing a button marked *Test* (or Self-Test or Slope Test).

The integrator measures the chromatographic signal from the detector for a set period of time (<1 min). It then examines these data for the largest change in slope during the test time and sets a *minimum slope value*. Any slope change greater than this value will be defined as a peak start or a peak end. Next, it will average all deflections over the test period and uses this average to set a *noise reject value*. Anything above this level will be treated as signal; anything below this level will be ignored. Finally, after an injection is made the integrator will use a preset *peak width* window and test to see if the current peak falls in this window. If it determines that the peak is wider than the window, the peak width will be automatically double for the next peak and this new value will be tested during its acquisition. This corrects for the tendency of peaks to widen, the longer they stay on the column. In gradients, where peaks may sharpen, the integrator can cut the peak width by half.

I have been told that an integrator should be able to integrate a difficult separation four times more accurately that you can integrate it manually. This is often not seen when you examine repeated injections made with machine set parameters. The Test button is useful when running first time or unknown samples and should be used when resetting detector sensitivities since these make major changes in baseline noise and slope values.

Be aware that when you press the Test button, you are allowing the computer to make a decision about minimum slope, noise, and peak doubling points for you. Once a separation has been run a few times, more precise data can usually be obtained by manually setting the peak width and noise values and by forcing peak doubling to occur at the specific points in the chromatogram using time programs built into many integrators.

14.4.2 Peak Detection and Integration

Once data are in the computer, they must be processed to determine the identity and relative amounts of each material present. First peaks must be defined. This is done by detecting peak starts and ends based on slope changes, determining *retention times* by detecting peak centers by inflection point changes from positive to negative slopes, and then calculating either peak heights or peak areas. Next relative areas or heights are calculated by summing all detected areas or heights, then dividing each individual value by the total for the chromatogram.

14.4.3 Quantiation: Internal/External Standards

Finally, these relative heights or areas are compared to equivalent values from standard curves prepared from known amounts of compounds with the same retention times to yield values for the amounts of each compound present.

The decision to use either peak or area data is based on the nature of the mixture analyzed. Fairly simple mixtures, such as reaction mixtures or fairly clean extracts, are usually more accurately determined using area data. Very complex mixtures from biological sources, such as blood or urine, are best quantitized using peak heights, since peaks are often detected on the shoulders of earlier, trailing peaks. This distinction is so common that you can usually separate a roomful of chromatographers into clinicians and researchers simply by asking who uses peak heights or who uses areas for quantitation.

Calibration standards can be of two types: external standards and internal standards. With external standards multiple concentrations of the standards are injected, areas are measured, and a calibration curve is plotted. Unknowns are then run, and areas are calculated and compared to the calibration curves to determine amounts of each compound present. With internal standards, known amounts of an internal standard are added to each known concentration of standard compounds and areas or peak heights response factors relative to those of the internal standard are calculated. When unknowns are run the same amount of internal standard is added to the unknown sample, and relative areas or heights are calculated based on the response factors to the internal standard from the calibration curves. Internal standards are usually used to correct for variation in injection size due to different operators and injection techniques. They can also be used to correct for extraction variation, but generally one or the other, not both at the same time.

14.5 Automated Methods Development

If you will pardon the pun, this is a developing field. There are no generally accepted approaches to methods development, but we will look at the underlying logic. Automated methods development is modeled after manual development of a standards separation. The instrument "looks" at a chromatogram, makes a change, and then "looks" at the new chromatogram to see if the separation has improved. Changes are made "systematically" until the "best" separation is reached.

14.5.1 Automated Isocratic Development

Obviously, to make this happen we have to (1) have a way of measuring the completeness of a separation, (2) define what constitutes the best separation, and (3) define a systematic pattern for making changes.

When we look at a separation to judge whether two peaks are separated, we look at the centers of the peak, but, more importantly, we look at the valley between the peaks. An ideal separation is one in which all peak pairs are baseline separated; the peak valleys all come down and touch the chromatographic baseline. If we were to draw a line connecting two peak tops (Fig. 14.2, line 1–2), then drop a perpendicular line from the center of this connecting line to the baseline (A–B), the length of the resulting line would represent a standard of baseline separation for the two peaks.

If the valley did not touch the baseline, it would have a length (A–C) that

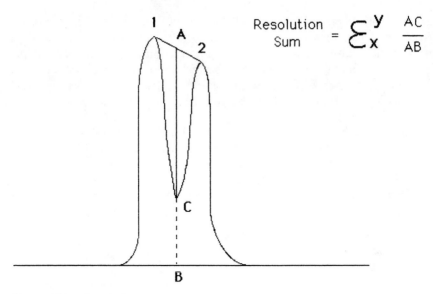

$$\text{Resolution Sum} = \sum_x^y \frac{AC}{AB}$$

Figure 14.2. Resolution sum.

would be less than the distance to the baseline. If we divided the length of A–C by the length of A–B, we have a mathematical measure of the baseline resolution for two peaks. Summing these for baseline resolutions for every pair of peaks gives us a *resolution sum* for the whole chromatogram.

Once we can put a relative value on the degree of baseline resolution between two peaks the rest is easy. The next step is to work to make the number as large as possible moving us toward a "better" separation. The computer changes one system variable, has the autoinjector make a new injection, and calculates a new resolution sum. Through repeated variable changes, new injections, and recalculation, it works to maximize this value.

Left to itself, the computer will carry this optimization to ridiculous extremes, so we must place search *limits* on the process. We would like the final chromatogram to run in a reasonable time, so we set a limit on the length of the *expected run time*. We also need to specify the *expected number of peaks* in this run time. Now the computer will search until it has maximized the baseline for the expected number of peaks in the preset run time.

The final information needed is a search pattern and the independent variables to be searched. Usual variables are flow rate, %B, and %C; %A is always assumed to be (100% − the sum of the other solvent percentages). One commercial search pattern starts with all variables at zero, then systematically changes one variable by a preset percentage and walks incrementally through all possible values, then repeats for the next variable. Once all injections and chromatograms have been run, each run is inspected and the best value is selected.

A second search pattern makes injections with each variable in turn set at zero with all others at maximum. It then makes injections with each pair of variables at half maximum and the remaining variables at zero. Finally, it makes an injection with all variables at half maximum and interpolates to predict the best separation. To visualize this technique, we would use three variables and place the 0% of each at the corners of a triangle. First, we would run compositional values at each corner, then the middle of each side, and, finally, the center of the triangle.

A more sophisticated method uses a random walk or simplex optimization search pattern. Variable limits are set, then three conditions within these limits are selected at random, injections are made, and chromatograms are run. The resolution sums for the injections are measures, the lowest value is discarded, and a new variable setting is selected directly opposite the discarded value and equidistant from a line connecting the two remaining values from the original triad (Fig. 14.3).

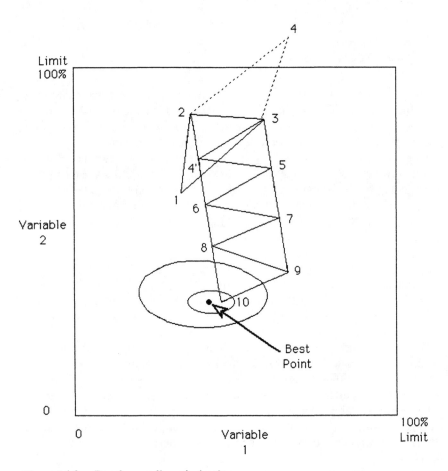

Figure 14.3. Random walk optimization.

If the new point falls outside the limits, it too is discarded and a new point is selected halfway between the first triad discard and the line equidistant between the other two points of the triad. Now, a new run is made using the variable conditions of this new point and the resolution sum is calculated. This process continues with the lowest value of each new triad being discarded, reflection around the axis joining the best two points and injection at the new set of conditions. This technique will hunt and search until a point is found that meets the search criteria. The search can be stopped at this point, but there is a danger that only a local "best" value has been found. If the overall best conditions are desired, this final point can be discarded, a new point selected at random, and the random walk can be continued. If the computer continues to return to the previous "best" point, then it probably represents the true best value within the limits. Obviously, a limiting maximum number of injections should be set to keep the computer from wandering around forever.

Although flow rate and solvent composition are the most commonly optimized variables, there is no reason why other mobile phase modifiers could not be used. Variables such as mobile phase pH, buffer concentration, ion pairing reagents, chelater concentration, or organic amine modifiers could all be optimized using resolution sums. If the variable wavelength UV detector can be controlled by the computer, wavelength and sensitive setting could both be included as independent variables to be searched and optimized.

To create a method, you would need a computer-controlled gradient HPLC system, with an autoinjector capable of making repeated injections from a large supply of the target solution, an HPLC column, and a detector. Data acquisition and processing can be done in an integrator and sent to the computer or can be handled directly by the computer. The system would be set up with sufficient mobile phase for an overnight run, limits set, and the system allowed to be run unattended overnight. When you come in the next day, the system will be either still running chromatograms or the report will be ready with the best chromatographic conditions on the final printout.

14.5.2 Hinge Point Gradient Development

The system is usually designed first to try to optimize a two-solvent isocratic separation (i.e., %B and flow rate). If this cannot be achieved within the run time and expected peak limits, a decision must be made as to the next type of development. If your system is a two-pump gradient system, the next step is probably development of a binary gradient. If you have a multisolvent gradient system, you usually try to optimize a three-solvent or perhaps even a four-solvent isocratic separation in the same fashion that we optimized a two-solvent isocratic separation.

To create a binary gradient, a linear gradient is run from 0 to 100%B, the resolution sum calculated, and then a hinge point development is begun, as discussed in Chapter 12. One hinge point at a time usually is selected and optimized. Peak pair resolutions can be examined to find areas of compacted peaks and holds established before the compaction areas to improve resolution sums. After resolution is maximized for compacted areas, slope increases can be introduced at random hinge points to speed run time while maintaining resolution. Gradient development is very much a research science at the moment.

If neither binary gradients nor three-solvent isocratics are successful, some systems will next try to perform a three-solvent gradient optimization. This development is very difficult to visualize. Assuming simultaneous optimization of %B, %C, and flow rate hinge points, it takes a long, computation-intensive time to carry out. It would be nearly impossible to carry out manually. The key is continually to use *the rule of one;* change only one variable at a time.

These last changes are probably of academic interest only. Most separation can be achieved with either a two-solvent isocratic or a binary gradient. Amines that tend to tail under neutral pH complicate the development. Moving to an endcapped column or adding a fixed amount of organic modifier will usually fix the problem. Acids can be handled by going to low pH using a fixed amount of acid to buffer pH.

14.6 Data Exportation to the Real World

Raw data and reports can be stored in the computer's memory, but they must be transmitted to the real world to be of use. In the simplest case, they can be displayed on the computer monitor in the form of chromatographic curves, tables of data, or reports, or they can be sent to the printer for printing. They can also be shared with other computers or with other software applications for further processing. To move data out of the resident software program, they generally have to be translated into some standard format recognized by other applications.

14.6.1 Word Processors: .ASC, .WS, .WP Formats

The simplest of the formats used to transfer data into word processing applications is the ASCII (.ASC) format. ASCII is a standard set of 128 binary codes used by all computers to represent all the characters present on the normal or shifted keyboard plus control codes originally intended for use on teletypewriters. These codes allow us to display small letters, capital letters, numbers, and punctuation marks on the computer screen. Other codes control line feeds and carriage returns, but formatting codes for underlining, boldface, and italics are not included in ASCII, and are removed in converting to this format. ASCII files have space-separated code and can be sent out over a modem or a serial cable to another computer and applications importing ASCII code. Other word-processing formats recognized by many other writing applications include the Word Star (.WS) file format and the Word Perfect (.WP) file format. Both of these are modified ASCII formats that transfer some formatting information with the file.

14.6.2 Spread Sheets: .DIF, .WK1, .WK2 Formats

The next type of standard output format is the spreadsheet. These file formats use comma-separated ASCII code, but also add addressing information for the columns and rows they occupy. The simplest are .DIF files, which originated to allow information transfer between VisiCalc worksheets in the Apple II computer and have been retained as a standard format. .WK1 and .WK2 are Lotus-1,2,3 formats that have come to be spreadsheet standards allowing transfer of data, addresses, and macros.

14.6.3 Data Bases: .DB2 Format

To export data from reports into a database program, a data base file format called .DB2 was developed in an early PC data base, dBase II. Data bases are made up of *files,* which could be compared to a rolodex box full of cards all containing the same type of information. The rolodex card would be equivalent to a data base *record.* Each record has on it a series of entries, *fields,* in the same place on each card. To import chromatographic data into a dBase record, all the entries in the report must match up with existing dBase fields.

14.6.4 Graphics: .PCX, .TIFF, .WPG Formats

Graphics, the fourth type of export from chromatographic data, is the most difficult. We can export copies of the monitor screen as bit maps in standard graphical formats such as .TIFF or .PCX files, but much of the fine detail and companion information will be lost. These bit map files can be manipulated, cleaned up, and labeled in "Paint"-type applications, and then imported into word-processing packages. However, the chromatogram can no longer be resized and data extraction or integration is no longer possible. Word Perfect graphic files for use in .WP text files can be exported by some applications. In some graphical applications it is possible to write a printer format to a file. The Plotter format (.HPLP) used by Hewlett Packard plotters is a Vector Graphic file similar to a Postscript file and can be used by some applications to resize, rotate, and reprocess the graphical output.

14.6.4 Chromatographic Files: Metafiles and NetCDF

Chromatographic data file formats are very often in system- and manufacturer-specific metafiles. The formats that are used to store these files within an integrator or data processing unit are usually not designed for export, or they are designed for export only to other modules by the same manufacturer. They may be in a proprietary format, in a compressed storage format, or even generated under a different computer operating system. Many offer the capability of translating part of their contents to ASCII, but a great deal of information, especially graphical information, is usually lost in the process.

To overcome this problem a standard chromatographic file format, NetCDF, has been approved by a committee of chromatographic companies. Similar formats are being developed for spectrographic and mass spectral data. In theory, the manufacturers and interested third-party software suppliers will provide translations of existing file formats to NetCDF standard format allowing file translation of data from one data system to any other system.

This would be a great boon to data security, but there is some question of its viability. The chromatographic standard has been in place since 1991, but there has been little rush to adopt it. As more and more data systems are declared obsolete and no longer supported by their suppliers, it may become obvious to research laboratories how transient and fragile their data files really are. This may lead to a rush to get them into a standard format.

Where Do We Go from Here?

We have already touched on some frontier application areas in HPLC: polysaccharides, larger mammalian polynucleotides, phospholipids, restriction fragments, DNA amplification reaction mixtures, and faster drug screens with positive compound identification. Clinical and environmental researchers are always pushing toward higher sensitivity detection and positive compound identification. What new equipment, columns, and techniques are already on the horizon?

15.1 Microsystems and Superdetectors

For some time there has been talk of "sealless" pumps to overcome the continuous problem of torn, worn out, and leaking seals. Companies periodically show up at meetings with flexing seal pumps and rumors circulate about a "Wankel" rotary pump with no plunger and floating seals.

Detectors are becoming more sensitive and less bothered by temperature and turbulence. UV detectors are already capable of reaching 100 pg sensitivity; fluorometers already exist that can reach 5 pg for selected compounds from clean matrices such as brain tissue. Solvent purity considerations and impurities in a complex matrix alone prevent them from achieving their maximum sensitivities without excessive noise. Refractive index detectors and conductivity detectors are reaching 50 ng sensitivity levels. Concentration correcting, scanning UV detectors are already available, and correcting, scanning fluorometers should be available soon. Eventually, we should see diode array detectors in which array chatter has been overcome, leading to much higher sensitivity. I also expect to see a "smart" diode array detector that scans, selects the best wavelength, checks for peak contamination (front, top, and back), and outputs information to a "smart" integrator allowing real time output of contamination annotated chromatograms and postrun peak printouts with information on the three strongest wavelength maxima/extinctions for each peak. Right now diode array detectors are used in real time detection mainly as glorified, expensive variable detectors with all information processing taking place in a batch mode long after the run has been completed. Eventually, diode array output will be summed like a mass spectrometer's total ion current (TIC) chromatogram for real time total spectra chromatogram (TSC) display. Postrun processing will involve single wavelength chromatogram (SWC) extraction and spectral extraction with comparison to spectral data bases for compound identification.

As computer hardware, memory, and software continue to plummet in price, look for reasonably priced data acquisition systems to become fully integrated with spreadsheet and data base programs, to produce systems that can acquire, convert, and store data automatically. It will be completely available for report generation with integral graphics or comparison to remote on-line data bases for compound identification.

15.2 LC/MS Primer

One of the most important potential additions to the HPLC arsenal is not a new technique. The mass spectrometer has been around for a long time with its major shift into the research laboratory as an outgrowth of the Manhattan Project during World War II. In the 1960s a GC/MS interface was developed, but the first HPLC/MS interface did not appear until the 1970s because of the problem of seeing compounds in the presence of all that solvent. The mass spectrometer is very nearly the perfect HPLC detector allowing noncontroversial identification of even "unknown" compounds from their fragmentation spectra. The problems preventing widespread introduction of LC/MS into general laboratory use have been twofold: price and expertise in interpreting results. Ten years ago a mass spectrometer system was a massive instrument costing in excess of $100,000 with an LC interface costing in excess of $20,000. The high vacuum pumps required to run the system needed constant maintenance. Interpretation of spectra was tricky and required someone specialized in the field.

Many of these problems are rapidly disappearing. Desktop mass spectrometry detectors (MSD) have shrunk in size and prices have dropped to around $48,000, pumping systems are becoming less service demanding, and interpretation of results much easier and faster due to computerized on-line, rapid spectral library data base searching.

While prices of new systems are still prohibitive for the average chromatography laboratory, older systems have been retrofitted with modern data systems, equipped with home built LC interfaces, and put back into operation for around $30,000. As prices continue to drop and technology advances, the LC/MS will become a major tool for the forensic chemist whose data must stand up in court, for the clinical and pharmaceutical chemist whose separations impact life and death, and for the food and environmental chemist whose efforts affect the food we eat, the water we drink, and the air we breathe.

With this in mind, let us take a look at the design of the LC/MS, its operation, and the way mass spectral data are manipulated to produce chromatographic information and compound identification. This will be simply an overview. Mass spectrometry is a field in itself, but it is important for the chromatographer to have a working knowledge of its techniques.

15.2.1 Quadrupole and Magnetic Sector Mass Selection

Mass spectrometers work on the principle that a charged ion being propelled through a curved magnetic field will be deflected inversely proportional to its molecular mass and proportionally to its charge. The lighter the mass the more deflection that will occur at a given charge. The higher the charge the more deflection that will occur at a given mass.

The earliest mass spectrometers, and many of the best systems currently used for determining accurate mass, employ large permanent magnets to establish the electromagnetic field (Fig. 15.1). The entire internal sample pathway is run under very high vacuum. The sample to be analyzed using a direct probe is placed into a "source" where it is ionized, repelled into the magnet, and focused by differentially charged electrical plates or grids called "lens." Masses are separated in the electromagnetic analyzer and detected when they impact at a specific point on a detector. Since some molecules can support multiple charges, the detected mass is usually referred to as m/z or the mass divided by the charge.

Quadrupole mass spectrometers also employ a source, lens to get the charged ion into the analyzer, and a detector all under high vacuum. However, a quadrupole electromagnetic field is applied to the mass analyzer rods and is modified with a radiofrequency signal for mass separation and to select and focus the desired mass at the detector (Fig. 15.2). By sweeping the Rf field through a range of frequencies, the quadrupole can be made to focus a series of differing mass ions on the detector allowing continuous measurement of m/z through a selected AMU range (SCAN mode). Alternatively, the quadrupole can be stepped to specific AMU values in a single ion monitoring (SIM) mode. Scan mode is generally more useful when doing qualitative detection and in fragmentation studies of unknowns. SIM mode is used for high sensitivity detection and quantitation.

One other commonly used mass spectrometer is the tandem mass unit also referred to as MS/MS or as a triple quad. This is actually two of three mass spectrometers used in series. One MS is used to fragment each compound, the

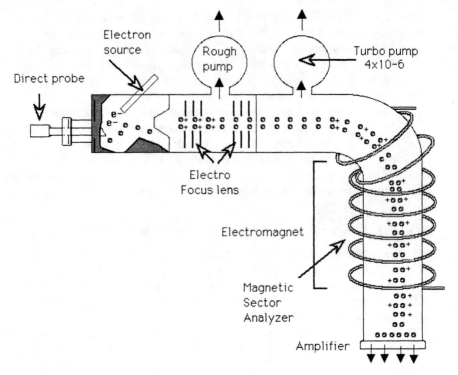

Figure 15.1. Magnetic sector mass spectrometer.

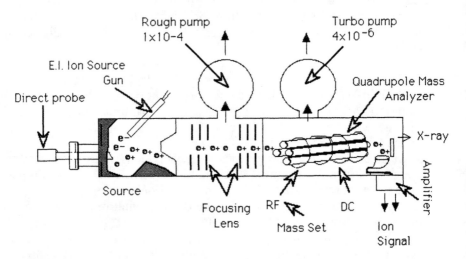

Figure 15.2. Quadrupole mass spectrometer.

middle unit allows further gas collision fragmenting and the third MS is used to look for a specific common ion. Alternatively, the first MS will select only single molecular ions, the middle unit allows fragmentation promoting collisions to occur, and the third quad is used to scan the fragmentation or look for specific ions from the fragmentation.

15.2.2 LC/MS Sources

The mass spectrometer source controls the nature of the sample reaching the analyzer. High-energy ions produced in *electron ionization (EI)* sources tend to reproducibly fragment into a cascade of charged and uncharged pieces that can be used to identify the original compound. A low energy source, such as a *chemical-induced (CI)* ionization source, can produce intact ions that can pass through the source into the analyzer without fragmentation to produce only molecular ions for molecular mass determination.

The problem faced in the LC interface is the introduction of larger volumes of the mobile phase along with the compound to be analyzed into the high vacuum environment of the MS source. LC/MS began in 1969, with a 1 μl/min flow into an EI source. In the 1970s, an LC source was developed using a continuous, moving metal band that pulled a portion of the column effluent into a heated vacuum oven. Flow rates were reduced by using a splitter in the effluent line.

The first modern *thermospray (TSI)* interface (Fig. 15.3), introduced in 1983, allowed introduction of column effluent at 1.0–1.5 ml/min. The mobile phase had to be highly aqueous and contain large amounts of volatile buffer to induce chemical ionization. The mobile phase was forced through an electrically heated capillary and out through a fine orifice into a high vacuum. Explosive last stage vacuum evaporation of the solvent droplet in the presence of high buffer chemically ionized the sample, which was pulled by a voltage differential into the analyzer.

The fact that the entrance capillary must often be headed to >200°C and the requirement for >100 mM volatile buffer often lead to sample decomposition and orifice plugging. Roughing vacuum pumps were added to remove the initially large amounts of evaporated solvent. An interface was developed that allowed solvent gradient operation by postcolumn addition of a compensating amount of high buffer solution.

Electrospray (ES) shows tremendous application for protein detection and analysis. ES is limited to microflow applications (1–50 μl/min) and involves effluent being forced through a capillary out into the source vacuum through a coronal electron discharge (Fig. 15.4).

Proteins can acquire multiple positive charges at basic amino acids such as lysine. Since the MS analyzer separates on the basis of m/z, or mass divided by charge, mass spectrometers with an operating range of 0–2000 amu can still detect proteins with 10–50 charges and masses up to 100,000 amu. Deconvo-

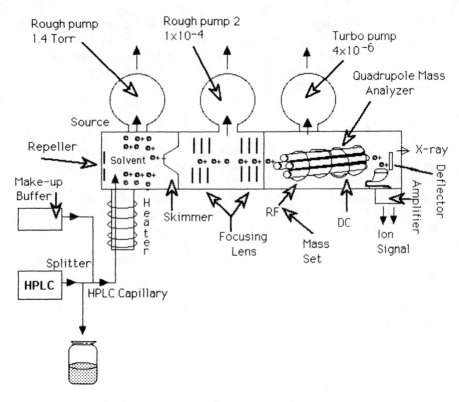

Figure 15.3.   Thermospray interface.

lution of the charge envelope developed by a protein allows calculation of the protein's molecular weight.

The *atmospheric pressure interfaces (API)* show real promise for general HPLC application. They employ an inert gas sleeve around the effluent inlet capillary to provide a high-velocity gas jet to produce a solvent mist to aid evaporation. In the heated *nebulizer interface (HNI)* makeup nitrogen sweeps the sample droplets into an electrically heated tube and then out over a discharge needle. Charged sample ions are pulled by a voltage potential difference through a curtain gas into the analyzer rods. The *ion spray interface (ISI)* reverses the process (Fig. 15.5). The nebulizer gas converts the effluent into a fine mist in the presence of a high electrical potential. Large droplets are removed by a grounded liquid shield and fine, charged mist is pulled into an electrically heated capillary in a first-stage vacuum chamber. The charged molecules are pulled into the analyzer and onto the detector.

Both interfaces allow high sensitivity HPLC operation at 1–1.5 ml/min flow rate without a stream splitter, run gradient effluents without makeup buffer, and produce a CI molecular ion. ISI has the advantage of being able to produce

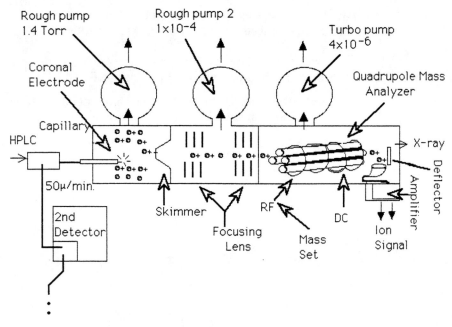

Figure 15.4. Electrospray interface with split.

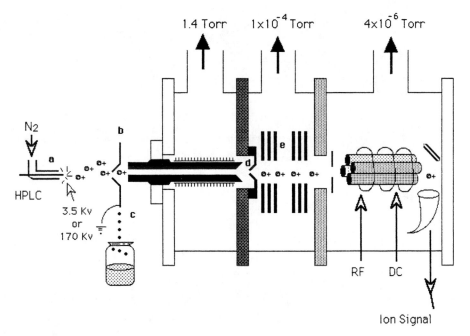

Figure 15.5. Ion spray interface.

EI-type fragmentation by increasing the voltage potential between the nebulizer tip and the liquid shield, effectively a poor man's MS/MS.

15.2.3 Chromatograms from the Mass Spectrometer

Data from the mass spectrometer are a three-dimensional block of information
similar to the output of a diode array UV detector. The data axes are time,

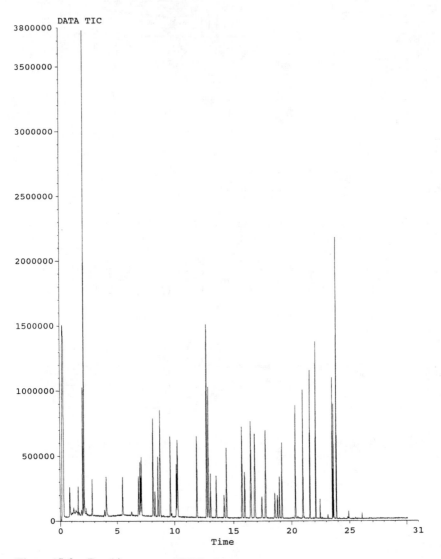

Figure 15.6. Total ion current (TIC) chromatogram.

intensity, and mass measured in amu. There is a difference, however, in the way a chromatogram is displayed in mass spectrometry.

The TIC chromatogram (Fig. 15.6) is a summation at each time point of all the mass intensities present.

A single ion chromatogram (SIC) (Fig. 15.7) is a chromatogram at a single mass and is usually extracted out of the TIC. It would be the equivalent of the normal UV chromatogram at a single wavelength.

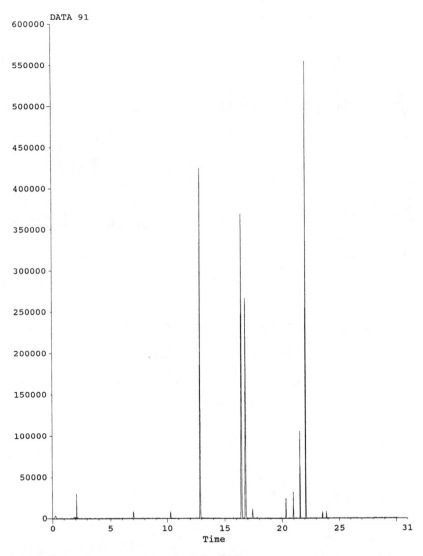

Figure 15.7. Single ion chromatogram (SIC).

A data slice at unit time is a mass spectra, equivalent to a UV spectra at unit time in a diode array detector. In a CI mass spectrogram, the spectra will be a single peak at the molecular ion molecular weight. In an EI mass spectrogram (Fig. 15.8), it will be a spectra of fragmentation ions that can be used to determine the molecular structure. Depending on the stability of the molecular ion, it may contain a peak equivalent to the molecular ion molecular weight.

Considerable expertise is required to do molecular structure calculations

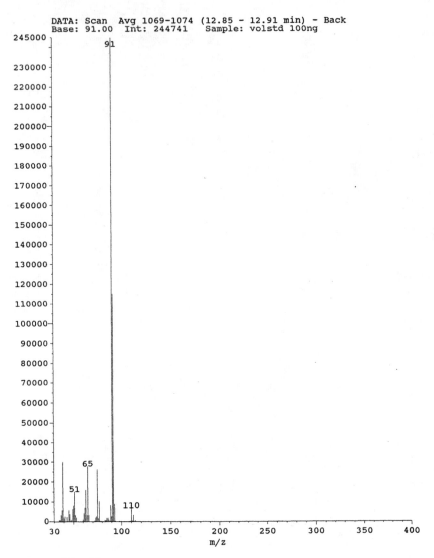

Figure 15.8.   EI spectra.

from fragmentation data. Fortunately, a number of libraries of mass spectra of known compounds are available and many can be computer searched to determine matches to known compounds. The most commonly used are the NIST Data Base Library (>64,000 compounds) used by many environmental laboratories and the Wiley Registry (120,000 compounds), which contains many more research-oriented compounds. Although the spectra in these data bases are of pure compounds, the search algorithms can often be altered to allow compound identifications in mixtures.

15.2.4 Automated Mass Chromatogram Quantitation

TIC and SIC can both be integrated with peak detection integration in a manner very similar to UV chromatograms. Internal and external standard quantitation can be made against calibration runs of known standards, but with a critical improvement over the usual UV quantitation. In target compound analysis, one mass or a summation of spectral masses can be used for quantitation while certain other masses can be used as qualifiers. A qualifier mass is a secondary mass fragment that must be present, and/or be present in a known ratio to the target mass before the major mass is used to identify the unknown peak as the correct compound. Since multiple qualifier peaks can be required, it is possible to use quantitative TIC information to produce definitive identification of a compound and the amounts of it present.

Application of similar chromatography display (TIC) and quantitation might allow real time display of meaningful information from a UV diode array detector. Acceptance of an equivalent target compound identification might go a long way toward making the diode array useful in definitive quantitation.

15.3 New Column Directions

Column packings have continuously gotten smaller, more homogeneous, and more regular. Column-to-column reproducibility within the top companies is always improving. The short 3-μm packing columns give the same separation as the 25-cm analytical columns in one-eighth the running times. Quick quantitative looks at two or three close running components may take only 1–2 min.

The trend from 10-, 5-, and 3-μm packings may go a bit farther when very tiny packings can be made homogeneous, but backpressure is going up rapidly. Even the best of these column materials is still a distribution of sizes around the stated packing diameter. Spherical packings of a uniform diameter are theoretically possible and should produce a higher plate column and more uniform running conditions. Look for wall-bonded brush phase in capillary columns to be a new major direction for HPLC and CZE.

Bonded-phase functionality has increased and begun to fill the gap between the C_8 and silica column. Expect to see more polar and hydrophilic function columns for normal phase, a phenol column, and a guanidine column, as interesting weak ion-exchange columns.

Size-separation columns will be extended down to cover small proteins and large and small peptides. An HPLC desalting size column is badly needed to remove buffers and salts from purified macromolecules. On the high end look for columns to separate RNA, DNA, and even suspended cell classes from each other. A monoclonal cell separator will be very important.

Affinity columns with substrates for specific enzymes attached to the column surface have been used for open column enzyme purification. Expect to see them available soon in kit form for HPLC; bind your own substrate to an activated bonded phase, pack it in the cartridge, compress it, and run it to remove reaction by-products.

15.4 Dedicated Systems: Pushbutton Clinical and the Personal HPLC

Systems have gotten simpler and less expensive on one end, while on the other they have become more expensive and elaborate. Quick release high-pressure fittings will outdate the tool bag. Simple, smart pumps will become interchangeable "disposable" items. Triplicate injectors will allow multiloop loading for repeat self-injections.

Two new system types will emerge. The first will be the dedicated, "clinical" analyzer: inexpensive, single analysis, isocratic, microbore, microflow, high-speed, multiinjection, with integral data processing/report generating. The works in a box. Ideally, this system may include a time of flight mass spectra detector as the final and ultimate identification tool.

The second will be the "full blown" personal research gradient HPLC: multisolvent inlet (at least three) dual headed pump, microautosampler with manual load bypass, column switcher with reinjection holding loops, multicolumn bank, double reagent postcolumn reactor, combination scanning diode array UV/fluorescence, and microprocessor-based data control system. The latter will handle data processing and communication and menu-based control of system parameters: pump speeds, gradients with display, solvent switching, column switching, wavelength scans, detector switching, and flow collector automation. All changes will be time stamped on the chromatogram along with operator identification and operating parameters for good laboratory practice. In addition, following the trend of microengineering in the computer field, the whole unit will take less than a square foot of bench space with stackable module clipping together and be fully portable running off a cigarette light outlet.

If this sounds far-fetched, imagine in 1978 describing a portable computer weighing 4 lbs with 8 mb memory, 250 mb of storage, a color screen, and all running on batteries!

Appendixes

I have included five items as part of the Appendixes. The first is a Separation Guide to point out starting points for chromatographic separations and also to point out trends in usage of columns, mobile phases, and detectors.

The second item is a glossary of HPLC terms. I have tried to include much of the terminology and buzzwords used in the field.

The third item is a troubleshooting quick reference. It is not intended to replace the systematic approach to troubleshooting in Chapter 10. When things go wrong, however, you may find it helpful. I have arranged it in the way things flow through the system; from pumps to integrator.

The fourth item is a series of three HPLC laboratory experiments. The first familiarizes the student with getting a system up and running and calibrating a column. The second experiment shows how to clean a column and pacify a system. The last is a first, quick-look at method development. These are three areas where I feel my students could best benefit from tools I have developed.

The last appendix is a selected reference list. It is not intended to be exhaustive; rather, it is simply to give you a point to enter the literature in the field. To stay current you probably want to at least subscribe to *LC/GC Magazine,* the *Journal of Liquid Chromatography,* and *Chemical Abstracts—HPLC Selects. Chemical Abstracts* are also on-line as part of Dialog's computer data base and can be searched through COMPUSERVE.

Personal Separation Guide

Application	Column	Detector	Conditions[a,b]
1. Vitamins (water soluble)	C_{18}	UV (254 nm)	8% AN/H_2O, C_7SO_3
2. Vitamins (fat soluble)	C_{18}	UV (280 nm)	80% AN/H_2O
3. Steroids	C_{18}	UV (230 nm)	60% $MeOH/H_2O$
4. Triglycerides	C_8	UV (220 nm)	60% AN/H_2O
5. Phospholipids	Si	UV (206 nm)	130/5/1.5—$AN/MeOH/85\%\ H_3PO_4$
6. Prostaglandins	C_{18}	UV (192 nm)	35% AN/H_2O, PO_4, pH 2.5
7. Bromphenacyl acids	C_{18}	UV (254 nm)	15–80% AN/H_2O
8. Krebs cycle acids	RNH2	UV (210 nm), RI	25–250 mM PO_4, pH 2.5
9. Monosaccharides	CX-Ca	UV (195 nm), RI	H_2O (80°C)
10. Polysaccharides	TSKpw	UV (195 nm), RI	H_2O (<20% AN)
11. Nucleic acids	CX-Na	UV (254 nm)	0.4 M NH_4HCO_2, pH 4.6
12. Nucleosides	C_{18}	UV (254 nm)	8% $MeOH/H_2O$, PO_4, pH 5.5
13. Nucleotides	C_{18}	UV (254 nm)	20% AN/H_2O, TBA, PO_4, pH 2.6
14. PTH amino acids	C_{18}	UV (254 nm)	10% THF/5 mM AcOH → 10% THF/AN
15. OPA amino acids	C_{18}	Fl (230/418 nm)	8% AN/PO_4, pH 2.6 → 3/25/30/40—$DMSO/MeOH/AN/H_2O$
16. Peptides (<99 amino acids)	C_8	UV (254 nm)	→30% n-BuOH/0.1% TFA, H_2O
17. Peptides	C_3	UV (210 nm)	40–70% AN/H_2O, PO_4, pH 5.5
18. Proteins (enzymes)	TSKsw	UV (254, 280 nm)	0.1 M Tris, PO_4, pH 7.0

Application	Column	Detector	Conditions[a,b]
19. Proteins (enzymes)	DEAE	UV (280 nm)	50 mM PO_4, pH 7.5 → +150 mM NaCl
20. Proteins (structure)	C_3	UV (280 nm)	0.1% TFA → 75% AN, 0.1% TFA
21. Catecholamines	C_{18}	UV (270 nm)	6% $MeOH/H_2O$, C_8SO_3, EDTA, pH 4.0
22. Theophylline	C_{18}	UV (270 nm)	7% AN/H_2O, PO_4, pH 4.0
23. Anticonvulsants	C_{18}	UV (220 nm)	40% $MeOH/H_2O$
24. Tricyclic antidepressants	C_{18}	UV (254 nm)	55% AN/H_2O, C_5SO_3, pH 5.5
25. Aspirin, Tylenol	C_{18}	UV (254 nm)	10% AN/H_2O, AcOH, pH 2.5
26. Aflatoxins	Si	UV (235 nm), Fl	6% MeOH/hexane
27. PNA	C_{18}	UV (254 nm)	80% AN/H_2O
28. Pesticides (carbamate)	C_{18}	UV (254 nm)	40% AN/H_2O
29. Pesticides (PO_4)	C_{18}	UV (192 nm), RI	50% $MeOH/H_2O$
30. Pesticides (Chlordane)	C_{18}	UV (220 nm)	80% AN/H_2O

[a] These separations are intended as a guide. They are not intended as recommended or standard procedures for *in vivo* diagnosis. Conditions will vary from compound to compound and from column to column.

[b] Abbreviations: AN, acetonitrile; AcOH, acetic acid; DMSO, dimethyl sulfoxide; PO_4, pH 2.6, phosphate buffer, pH 2.6; TFA, trifluoroacetic acid; C_7SO_3, heptane sulfonate; TBA, tertiary butyl amine.

APPENDIX
B

A Glossary of HPLC Terms

Adsorption Chromatography—Separation mode resulting from compounds have different adhesion rates for the packing surface (*see* Normal-Phase Chromatography).

Alpha (α)—(Separation or chemistry factor.) A measure of separation between two peak maxima. Ratio of their k' values.

Attenuation—Measure of detector sensitivity. A larger value means less sensitivity.

Autoinjector—An injection device for automated methods development in which the sample loop is repeatedly filled from a large sample reservoir rather that a sample vial carousel.

Autosampler—A multiple sample injector, usually with rack or carousel to hold sample vials, designed for unattended programmed operation in which a sample is loaded by either pushing or pulling sample into the injection loop with air or hydraulic pressure.

Autozero—Detector or integrator function capable of setting present detector signal value (baseline?) to zero.

Band—The disk of resolved compound moving down the column. Band spreading caused by diffusion tends to remix already separated bands.

Baseline—Detector signal versus time if no peaks are present. Good indicator of pulsing, air bubbles, electrical noise, or impurities.

Baseline Resolution—Chromatographic goal of methods development in which all valleys between adjacent peaks touch the baseline indicating complete resolution of peaks.

Buffer—Mobile phase modifier used to control pH. Usually salts of weak acids/bases, most effective at their pK_a, where concentrations of ionized and un-ionized form are equal.

C_{18}—Nonpolar column or packing with an 18 carbon hydrophobic hydrocarbon chain bound to silica through a silicone bonding. Used for reverse-phase separations. (See ODS.)

Cartridge Column—Disposable off-line tube packed with > 1 g of packing and used for sample and solvent preparation (*see* SFE and Windowing).

Check Valve—Mechanism in the pump head inlet and outlet to ensure one-way solvent flow. Usually a sapphire ball in a stainless steel cone. Major point of buffer precipitation and pump pressure loss.

Chromatography—A separation technique producing a qualitative record of the relative amounts of the components, a chromatogram. HPLC modes include partition and adsorption (polarity), GPC or SEC (size), ion exchange (charge), and affinity (specific retention).

Column—A metal tube in which the HPLC separation occurs, packed with porous packing held in place at each end by a fritted filter in an endcap. Endcaps are secured to the column with ferrules and can be opened for frit cleaning.

Column Bridge—A length of tubing, fitted with compression fittings simulating column ends, used to replace the column for system cleaning and diagnosis.

Compression Fitting—A device for connecting tubing to other system parts. Usually made up of a ferrule and a threaded screw or cap, which slide over the tubing. Tightening the screw/cap forces the ferrule into a conical hole squeezing (swaging) it permanently onto the tubing.

Dead Volume—Unnecessary volume in a system that can remix separated bands of compounds, usually in tubing or fittings, especially from injector to column and column to detector.

Deoxygenation—Removing oxygen from a solvent by vacuum replacement with nitrogen or helium to prevent oxidation of sensitive compounds or columns (amino).

Efficiency (N)—A measure of the narrowness of elution bands, the sharpness of peaks, and the performance of a column. Results are in theoretical plates. The Huber equation calculates efficiency versus flow rate, which is plotted on a Van Deampter plot.

Elution—Washing bands of separated materials out of the column with mobile phase. The liquid output of a column is the eluant; the amount of solvent needed to reach a peak's maximum is its elution volume.

Elutotropic Series—Solvents ranked in order of polarity or eluting strength. The strongest solvent is the one most like the packing material in polarity.

End Absorption—UV absorption, from 210 nm down, going nonlinear at 180 nm due to dissolved oxygen. Most carbon–oxygen containing compounds absorb in this area.

Endcapping—After silylation, reaction of the bonded-phase packing with a reactive small molecule to tie up unreacted silanols on the silica surface. Sharpens peaks from basic compounds.

Exclusion Volume—In size- exclusion chromatography, V_0, the volume of solvent necessary to wash out unretarded compound too large to penetrate the pores of a size-separation column. The inclusion volume, $2V_0$, is the elution volume needed to elute all compounds small enough to fully penetrate the pores.

Fines—Small particles of packing material in a column that tend to migrate and plug the outlet frit raising column backpressure.

Flow Cell—Low volume (8–20 μl) detector cell designed to accept eluant output from an HPLC or IC.

Frits—Porous stainless steel filters at either end of the column that serve as bed supports and filter the sample coming in from the injector.

GPC (Gel Permeation Chromatography)—Separation mode based on the molecular sizes of the compounds (*see* SEC).

Gradient—A reproducible change in a separation parameter that can be used to speed a separation. In a binary solvent gradient, % solvent B increases while %A decreases causing late eluting peaks to come off faster and sharper.

Guard Columns—Short, protective columns placed in-line between the injector and the main column.

Helium Sparging—A solvent degassing technique in which helium is bubbled through solvents to displace dissolved gases before solvent mixing, compression, and pumping.

Injector Seal—Hardened Teflon® plate within injector body used to separate system flow from sample loop until an injection is made.

Ion Displacement—Use of strong salt solutions to displace compounds bound to ion-exchange columns.

Ion-Exchange Chromatography—Separation mode for separating ionized compounds on charged columns. Anion-exchange columns attract and separate anions, cation-exchange columns separate cations.

Isocratic—Constant mobile phase composition. The opposite is a gradient in which the mobile phase composition is altered during the run. Isocratic conditions can include multiple components in the mobile phase.

k′—(Retention factor.) A measure of the relative solvent volume needed to wash a compound off a column for a given solvent polarity. Normalized with the void volume of the column to make it independent of column length.

Lamps—Light source for a detector. Deuterium is fully variable from 190 to 700 nm. Other lamps show discrete bands: mercury, 254 and 436 nm; cadmium, 228 nm; zinc, 214 nm.

Loop and Valve Injector—Device for placing sample onto the column head. Modern design consisting of a loop, partially or overfilled at atmospheric pressure, which is rotated into the flowing stream from pump to column. Sample is back-filled from the end of the loop closest to the column described as "last in, first out" filling.

Microporous packing—Modern fully porous, high-resolution separations packing with average particle diameters of 3–10 μm.

Mobile Phase—The solvent mixture pumped through the column carrying the injected sample. The liquid phase of the solid–liquid equilibration.

Needle Seal—Teflon® throat seal in injector needle port that prevents flowback of injected sample solution.

Normal-Phase Chromatography—Separations mode run on unbonded, anhydrous silica (polar) using a nonpolar mobile phase (*see* Adsorption Chromatography).

ODS—Octadecylsilyl bonded phase material or column in which the material bound to silica is an 18-carbon saturated hydrocarbon chain (*also see* C_{18}, RP_{18}).

Pacification—Treatment of a column bridged HPLC system with 20% (6 N) nitric acid to remove buffer and organic deposits and protect metal surfaces from corrosion. The column must be removed before acid treatment. Overnight water wash is needed to remove last traces of acid.

Peak Areas versus Peak Heights—Integration and quantization can be based on either the height or area of the peak. With well-resolved peaks seen in research laboratories, areas give more accurate results; with less well-resolved peaks or shoulders seen in clinical or biomatrix separations, peak heights give best results.

Pellicular Packing—First analytical packing. It had a solid core and a crust of porous silica. Now used primarily for packing guard column.

Plate Count—A measure of column efficiency derived by comparing peak width to retention time; higher numbers indicate a more efficient separation. Theoretical plates is an arbitrary unit assigned to the efficiency value, in analogy to efficiency units in open column distillations.

Plunger—A piston, usually of sapphire or beryl, driven by the pump motor into the pumping chamber to pressurize and displace solvent through the outlet check valve. The rear of the chamber is sealed by the *Plunger Seal,* made of hardened Teflon®, which fits tightly around the plunger.

Polarity—A measure of a solvent, column, or compound's ability to attract similar molecules. Polar compounds have large dipole moments, large dielectric constants, and usually form hydrogen bonds (e.g., water). Nonpolar compounds such as hexane are on the opposite end of the polarity scale (*see* Elutotropic Series).

Pulse Dampeners—Device used to control pump pulsing. Usually a tight coil of metal tubing that acts as a baffle and counters pump pulsing by a spring recoil effect.

Reciprocating Pumps—Single and dual headed pumps that use a piston and check valves to pump solvent from a reservoir into the system.

Resolution (R)—A measure of the completeness of a separation. Influenced by k' (solvent polarity), N (column efficiency), and α (system chemistry).

Retention Time—The time or mobile phase volume needed to elute and measure a component of the mixture in a detector.

Reverse-Phase Chromatography—Separation mode on bonded phase columns in which the solvent/column polarities are the opposite of normal-phase separations. Nonpolar columns require polar solvents.

Rotor Seal—Teflon® surface that seals the injector and separates the flowing mobile phase from the sample loop until an injection is completed.

RP_{18}—Reverse phase, bonded packing with 18-carbon side chain (see C_{18}, ODS).

Sample Clarification—Removal of particulates from the injection sample by either filtration or centrifugation.

Saturation Columns—Sacrificial column placed before the injector to protect the main column from pH degradation.

Seal—Wear surface that both lubricates and separates moving parts in the HPLC (*see* Plunger Seal, Rotor Seal, and Needle Seal).

SEC (Size Elution Chromatography)—A separation mode employing controlled pore size packing to achieve resolution of molecules based on size and shape (see GPC).

Separation Factor (α)—A measure of peak separation between peaks. Product of dividing one k' by the other.

SFE—Sample filtration and extraction cartridge columns used to filter and clean up sample before injection into the HPLC. SFEs with all modes of separation are available (*see* cartridge column).

Silica—Particles or spheres of crystalline silicic acid used in chromatography. Its surface is polar, acidic, and tends to attract water of hydration.

Silylation—The first step in forming bonded-phase packings from dried silica and dialkyl chlorosilanes.

Stationary Phase—A term used to describe the column packing indicating that it is part of a two-phase equilibrium with the mobile phase or column solvent.

Syringe Pump—A pulseless pump made up of a motor-driven piston in a solvent-filled cylinder. Useful only when small solvent volumes are to be pumped.

Tailing—Unsymmetrical peak formation in which the side of the peak away from the injection returns very slowly to the baseline. Usually due to an unresolved equilibration.

Voids—Spaces or openings in the column bed leading to poor chromatography. End voids are directly under the inlet frit. Center voids are channels through the center of the packing bed.

Void Volume—The solvent volume inside the packed column. Can usually be measured as an early refractive index baseline upset when injecting a sample dissolved in a solvent even slightly different from mobile phase.

Windowing—A technique using cartridge columns to speed chromatography by first removing polar and nonpolar impurities leaving only a solvent fraction containing compounds of interest.

Zero Dead Volume—Fittings designed to leave no extra column volumes that might cause band spreading or remixing of peaks.

HPLC Troubleshooting Quick Reference

This section is designed to assist in troubleshooting system problems. It is *not* meant to replace the systematic troubleshooting procedure in Chapter 10. A systematic approach is always better. Keeping this in mind, I have listed a series of commonly seen problems, possible causes, and suggested treatments.

Problem 1: No power. (Display does not light up on module or system.)
  Cause a: Not plugged in.
    Treatment: Check the plug at the socket and at the module. It may have been unplugged accidentally.
  Cause b: No fuse or incorrect fuse.
    Treatment: On a new system, make sure fuse(s) were installed. New systems are often shipped with fuses in a bag. If fuse is broken, replace it and contact service.
  Cause c: Not switched on.
    Treatment: Turn on switch (try the upper right-hand side of back first). Someone may have been helping you conserve energy. I always set my system up with a surge protector with an on/off switch. That way I can turn everything on with one switch and protect against line surges and ground loops at the same time.

Problem 2: Leaking fittings, puddles on desk top, fountains.
  Cause a: Compression fittings not tight enough.
    Treatment: Tighten leaking fitting another $\frac{1}{4}$ turn or until leak stops. (Leak will be at back of fitting around tubing).
  Cause b: Incorrectly made fitting or wrong ferrule.

Treatment: Stop pump. Loosen fitting with a wrench. Examine for correct preparation. Cut off and replace if necessary.

Cause c: Fitting scored by silica packing.

Treatment 1: Wrap fitting with Teflon® tape and reseal.

Treatment 2: Cut off ferrule and replace fitting.

Problem 3: Inconsistent or too slow pump flow rate.

Cause a: Air bubbles in pump head.

Treatment 1: Open a purge valve. Prime the pump with degassed solvent.

Treatment 2: Open the compression fitting at the top of the outlet check valve with a slow pump flow rate until solvent leaks around the fitting. Tap the side of the check valve with a small wrench until small air bubbles come out with the liquid. Reseat the compression fitting.

Treatment 3: Pacify the pump with 20% nitric acid after first removing the column. Wash with water repeatedly.

Cause b: Plugged solvent sinker in reservoir.

Treatment: Replace filter (Sinker is plugged. Pump is starving.) Try sonicating stainless steel frits in 20% HNO_3, then sonicate twice in fresh water.

Cause c: Outlet check valve is sticking open.

Treatment: Pacify with 20% HNO3 (see 3b).

Cause d: Sticky inlet check valve. Solvent flow back.

Treatment 1: Replace check valve.

Treatment 2: Pacify system with 20% HNO_3 (see 3b).

Problem 4: Pump pressure shuts down pumping.

Cause a: Overpressure setting set too low.

Treatment: Reset overpressure setting to a higher value, if column can tolerate it.

Cause b: Underpressure setting too low. No solvent.

Treatment: Reset underpressure setting higher. Check solvent reservoir and add more solvent, if needed.

Cause c: Column or system is plugged.

Treatment 1: If pressure exceeds 4000 psi, find the plug and clear it. Remove the column and run pump. If pressure persists, trace pressure back toward pump until the pressure drops. Reverse line and use pump pressure to blow out plug.

Treatment 2: If pressure leaves with column, replace the inlet frit (see Chapters 6 and 10).

Problem 5: No peaks detected after injection.

Cause a: No sample in syringe.

Treatment: After drawing up sample, pull back on syringe barrel so you can see the meniscus. Tap out any bubbles. Push syringe back to injection point. Wipe off excess liquid with Kimwipe. Inject.

Cause b: Sensitivity too low. Wrong wavelength.

Treatment: Check detector settings. Increase sensitivity. Inject again.

Cause c: Injector loop plugged.
   Treatment: Make sure solvent flows from injector waste line when loading
       loop. If not, disconnect injection loop at point closest to column and
       place injector in inject position with pump flow on. Wash loop into
       beaker and reconnect.
Cause d: Wrong elution solvent used. No gradient.
   Treatment: Use correct elution conditions. Make sure you started gradient
       run.
Cause e: Column has bound impurities.
   Treatment: Wash column with strong solvent. Equilibrate. Inject.

Problem 6: Injector leaks around body or around needle.
Cause a: Body leaks—torn injector seal.
   Treatment: Rebuild injector replacing injector seal.
Cause b: Needle seal—scored needle seal.
   Treatment: Tighten sleeve around needle seal or replace the seal.

Problem 7: Ghosting peaks occur when injecting solvent.
Cause a: Dirty sample loop.
   Treatment: Clean sample loop with a strong solvent.
Cause b: Small rotor seal tear trapping sample.
   Treatment: Replace the rotor seal.

Problem 8: Increasing column pressure.
Cause a: Column inlet frit is plugged.
   Treatment 1: Filter samples before injections.
   Treatment 2: Wash column with water before switching from buffer to
       organic solvent or vice versa.
   Treatment 3: Remove frit and replace it. Wash old frit by sonicating with
       20% HNO_3, H_2O, H_2O (see 3b).
Cause b: Column bed is plugged.
   Treatment: Do not inject a saturated solution. (Column will overconcen-
       trate and precipitate sample.) Wash out at best flow rate with strong
       solvent.
Cause c: Replace outlet frit. Sonicate old frit with 10% NaOH, H_2O, H_2O (see
       3b).

Problem 9: Column retention time and plate counts changing.
Cause: Bound material on column or voiding.
   Treatment: See Chapter 6.

Problem 10: Loss of detector sensitivity and dynamic range.
Cause a: Old detector lamp needs replacing.
   Treatment: Replace detector lamp. Record intensity value, if available, for
       later reference.
Cause b: Dirty flow cell window.
   Treatment: Clean windows. First, try cleaning in situ. Disconnect detector
       from system. Wash with water. Push 20% HNO_3 in from the waste

line and trap in flow cell. Leave for 15 min. Flush out copiously with water. If necessary, disassemble flow cells and clean window with acetonitrile, chloroform, and hexane using a Kimwipe. Dry and reassemble.

Problem 11: Baseline increases—no solvent flow/lamp on.
Cause: Decomposing coating on flow cell windows.
Treatment: Wash windows or pacify with 20% HNO_3. Disconnect column and wash into a beaker with strong solvent.

Problem 12: Rising and falling baseline.
Cause a: Late running peaks still eluting.
Treatment: Wash column with stronger solvent. Equilibrate with fresh mobile phase.
Cause b: Detector warm up is not complete.
Treatment: Go have another cup of coffee before shooting the next sample. If warm-up time becomes excessive consider replacing the lamp or calling a serviceman.

Problem 13: Noisy baseline or baseline spikes.
Cause a: Bubbles in flow cell.
Treatment 1: Add 40–70 psi backpressure device to detector outlet line.
Treatment 2: Disconnect detector. Push HNO_3 in to flow cell from outlet and trap in detector for 15 min. Wash with water.
Cause b: Electronic noise from power line.
Treatment 1: Get a line noise filter.
Treatment 2: Make sure that the detector signal line is properly shielded. Connect only one end of the shield line.

Problem 14: Peaks have plateaus or unexpected shoulders.
Cause: Strip chart slide wire is dirty.
Treatment: Pen is sticking. Wipe slide bar with Kimwipe. If necessary spray a little WD40 on Kimwipe and wipe bar.

Problem 15: Retention times vary in both directions.
Cause a: Stretched tension spring on chart drive.
Treatment 1: Check chart speed with stop watch. Reposition spring attachment to increase tension.
Treatment 2: Get a new integrator. Put strip chart back on GC or trash it.

Problem 16: Continuous retention time printing.
Cause: Noise level set too low. Too many peaks.
Treatment: Increase noise level value until only retention times of peaks of interest are printed.

Problem 17: Integration start/stop marks occur too early or late.
Cause: Peak width too small. Slope too high.
Treatment: Use autozero, test button before making injection. Recalculate using lower slope value or wider initial peak width.

D

Laboratory Experiments

The following laboratory experiments have been designed to let you try out the tools you need to run an HPLC system and its columns on a daily basis. In Laboratory 1, you will practice starting up an HPLC, recovering a dry column, and quality controlling a new column. In Laboratory 2, you will run a scouting gradient to select an isocratic condition. An SFE cartridge column will be used to window out peaks in the chromatogram. In Laboratory 3, we will look at the effect of changing the stronger solvent and changing column types on our separation. Finally, we will *remove the column* and pacify the system with 20% nitric acid followed by a water washout. Using these tools on a regular basis should keep your columns and systems up and running and provide procedures when you have to develop new separations.

Laboratory 1—System Start-up and Column QC

Purpose

1. To start up an HPLC protecting seals, plungers, and a "dry" column.
2. To run column standards.
3. To calculate plate counts/retention times.

Equipment and Reagents

1. Isocratic HPLC system
2. C_{18} column (5 μm, 15–25 cm)

3. 25-μl injection syringe
4. HPLC grade methanol and water
5. Column standards (P. J. Cobert Cat. No. 962202)
6. Column bridge (5 ft of 0.010-in. tubing, fittings, unions)
7. Backpressure device on the detector outlet

Protocol

1. Prepare 200 ml of 50% methanol/water and 100 ml of 70% methanol/water. Vacuum filter through 0.54-μm filter.
2. Remove the C_{18} column and set it aside. Set up the HPLC system with a column bridge in place of the column. Prime the pump(s) with 50% methanol/water. Set over pressure setting on pump at 4000 psi. Start flow at 0.1 ml/min and slowly ramp to 1 ml/min. Watch the pump pressure indicator for fluctuations (air bubbles? dirty check valves?). (*Lab Note:* Air bubbles can be cleared by opening the compression fitting on the outlet check valve with a wrench until solvent bubbles out, then tapping the valve house lightly to release bubbles. Retighten the compression fitting). Stair stepping pressure may indicate a dirty check valve, which should be pacified (see Laboratory 3).
3. When the pressure is steady, turn the injector handle to inject (or load if it already is in inject position) and watch pressure. If the pressure does not jump up, the loop is not blocked. Cycle the injector handle.
4. Watch the recorder baseline. When it is stable, slow the pump flow to 0.1 ml/min, remove the column bridge, and connect the C_{18} column to the injector. *Do not* connect the column to the detector yet. Wash it into a beaker (0.1–1 ml/min, slow flow ramp up) for 6 column volumes (12–18 ml). Pressure should slowly increase to around 2000 psi at 1 ml/min due to column backpressure. (*Lab Note:* Always hook up a column with solvent running to prevent introducing air into the column.)
5. When pressure is stable, *record column backpressure* from the pump pressure gauge. Connect the column to the detector. Turn on the detector (254 nm, 1.0 AUFS) and the recorder at 0.5 cm/min chart speed. Observe the basline. Drifting indicates that the detector is still warming up. (*Lab Note:* The pump pressure gauge should always be monitored when making system changes. A sudden pressure increase indicates a blockage problem. Adjusting the pump overpressure setting should prevent problems, but shut off flow yourself to be sure.)
6. When the baseline is stable, inject 15 μl of column standards. (*Lab Note:* Inject by overfilling the syringe, pull barrel back until you can see the meniscus, point needle up, tap out air bubbles, push plunger to 15 μl mark, wipe outside of barrel with a lab wipe with a pulling motion. Load the injector loop slowly, leave needle in place. Turn the injection handle quickly.)
7. On chromatogram, *mark the inject point. Record date, time, your name(s), flow rate, mobile phase, sample type and amount, detector wavelength and*

attenuation, and chart speed so you could duplicate this run. Record until baseline is reached after four peaks.

8. Repeat standards run. Increase recorder speed to 2 ml/min. Inject standards. Record the four-peak chromatogram.

Results

9. *Measure V_0, the exclusion volume of the column, V_x for each peak (the solvent volume at the peak center), and W for peaks 1 and 4.*

10. *Calculate k' (peaks 1 and 4), α (peaks 1,2), and N (peaks 1 and 4). (Lab Note:* Remember $k'(1) = V_1 - V_0/V_0$, $\alpha(1,2) = k'_2/k'_1$, $N_1 = 16(V_1/W_1)$. Also remember that W_1 is measured by projecting lines parallel to the sides of the peak to where they intersect the baseline. W_1 is the distance between the intersection points).*

Laboratory 2—Sample Preparation and Method Development

Purpose

1. Run a scouting gradient.
2. Select SFE cartridge column windowing conditions from the gradient.
3. Run SFE window cuts in selected dial-a-mix isocratic conditions.

Equipment and Reagents

1. Gradient HPLC system
2. C_{18} HPLC column (5 μm, 15–25 cm)
3. C_{18} SFE cartridge columns (Whatman Part No. 6804-0405)
4. 5-ml B-D disposable syringes
5. Test mix with seven-component mixture (P. J. Cobert Cat. No. 962201)
6. HPLC grade methanol and water
7. 4 test tubes

Protocol

1. Purge pump A with water and pump B with methanol. Dial-a-mix 20% B and equilibrate the column at 1.5 ml/min (6 column volumes or a stable baseline, about 10 min). (*Lab Note:* If the gradient system is a low-pressure mixing system, solvents must be degassed by purging with helium).

2. Inject 15 μl of the 7-component standard. Run a 15-min gradient to 100%, hold at 100% for 5 min. Watch the chromatogram during the run and *record %B of first peak and last peak.*

3. For a 25-cm column, deduct 10% from the first peak %B and equilibrate the column with this mobile phase (i.e., if the first peak came off at 80%, dial-a-mix 70%). For a 15-cm column, deduct 7% from the first peak %B. Equilibrate the column with this mobile phase.
4. Inject standards and run chromatogram.
5. Pretreat C_{18} cartridge columns with 2 ml MeOH, then 2 ml H_2O. (*Lab Note:* Remove plunger from 10-ml syringe. Put cartridge column on luer end of syringe. Put 2 ml of MeOH in syringe barrel. Push solvent and air through the cartridge with the plunger and collect eluant in test tube 1. *Do not pull back on syringe barrel while cartridge is on syringe.* Remove cartridge. Pull out plunger. Replace cartridge. Go to next solvent.)
6. Put sample in syringe, insert barrel, and push sample into the cartridge. Based on the scouting gradient, select three washes to window off standards: Select a window that will leave the three middle compounds in the second cut. [*Lab Note:* In windowing from a scouting gradient, start at the injection mark and move to the last peak you want to wash off. Find its equivalent %B on the gradient trace. Deduct 7–10% to find its isocratic equivalent for your first wash conditions. Collect in tube 2. Move to the last peak you want in your second (window) fraction and determine its wash off %B. Collect in tube 3. Finally, wash off remaining peaks with a strong solvent (i.e., 100% MeOH) for your final cut.] Collect in tube 4.
7. Run each wash out cut in the final isocratic mobile phase you selected for your HPLC separation.

Results

8. Examine the four chromatograms of the windowing cuts and eluant. *Measure retention times of the peaks in each cut and compare to the final methods development chromatogram.* If you have more or less than three peaks in cut 2, or if some of the cut 2 peaks are found in cuts 1 or 3, *indicate how you would adjust the %Bs of the window frame to improve cut 2. Do not repeat* the windowing experiment to prove your point.

Laboratory 3—Column and Solvent Switching and Pacification

Purpose

1. Study the effect of changing the stronger solvent or the column on the separation.
2. Do a column washout and QC check.
3. Do a system pacification using a column bridge.

Equipment and Reagents

1. Gradient HPLC system
2. Backpressure device on the detector outlet
3. C_{18} column (5 μm, 15–25 cm)
4. C_8 column (5 μm, 15–25 cm)
5. Column bridge (5 ft of 0.010-in. tubing, fittings, unions)
6. Four-component column standards (P. J. Cobert Cat. No. 962202)
7. Test mix with seven-component mixture (P. J. Cobert Cat. No. 962201)
8. HPLC grade acetonitrile, methanol, and water
9. Concentrated nitric acid

Protocol

1. Purge line A with water, line B with MeOH. Dial-a-mix 70% MeOH and equilibrate the C_{18} column at 1.0 ml/min. When stable, inject 15 μl of the seven-component test mixture and annotate the chromatogram. Run an isocratic chromatogram.
2. Reduce flow to 0.1 ml/min, remove the C_{18} column, and replace with a C_8 column (do not connect to detector). Increase flow to 1.2 ml/min and wash with 6 column volumes. Connect the column to the detector and run until the baseline is flat.
3. Inject 15 μl of the seven-component test mixture. Annotate and run isocratic chromatogram.
4. Dial-a-mix 60% MeOH/water. Equilibrate the C_8 column. Inject 15 μl of the seven-component test mixture. Annotate and run isocratic chromatogram.
5. Put acetonitrile in the B reservoir. Purge the line with acetonitrile. Dial-a-mix 60% acetonitrile/water. Reconnect the C_{18} column at 0.1 ml/min flow. Increase flow to 1.0 ml/min and equilibrate column. Inject 15 μl of seven-component column test mixture. Annotate and run chromatogram.
6. Make up 100 ml of 20% nitric acid (1 part acid to 4 parts water).
7. *Very Important Lab Note: Remove the column.* Replace the column with the column bridge. Put end fittings in the column for storage. Put water in the reservoir and wash the system at 2 ml/min with water.
8. Replace the water in the solvent reservoir with 20% nitric acid. Stop! *Note: Make sure the column has been replaced with a column bridge. Do not pump nitric acid through a column.* Wash system for 15 min at 2 ml/min with 6 N (20%) nitric acid. Discard the wash carefully.
9. Wash the system with water (2 ml/min) (set UV detector at 230 nm, 2.0 AUFS) and monitor disappearance of nitric acid. When baseline is flat or classtime has elapsed, collect effluent and check pH against laboratory water.

Results

10. Comparison of standards run on C_{18} and C_8 columns: For the gradient system, *compare C_{18} and C_8 chromatograms* (in both mobile phases). *Measure last peak retention times* in all three chromatograms. *Look for peak switching* based on peak heights.
11. Pacification of a columnless HPLC with nitric acid: *observe the length of time necessary to wash all nitric acid from HPLC system.*

Journals

1. *LC-GC Magazine*
   —"Sample Preparation Perspectives," Ron Majors
   —"LC Trouble shooting," John Dolan
   —"The Data File," Glen Ouchi

2. *Journal of Liquid Chromatography*

3. *Analytical Chemistry*

4. *Journal of American Society of Mass Spectrometry*

Papers

1. "Atmospheric Pressure Ionization—Mass Spectrometry Detection for Liquid Chromatography and Capillary Electrophoresis." Mark H. Allen and Bori I. Shushan. *LC-GC*, 11(2), 112–126, 1993.

2. "High-Flow Ion Spray Liquid Chromatography/Mass Spectrometry." G. Hopfgartner, T. Wachs, K. Bean, and J. Henion. *Anal. Chem.*, 65, 439–446, 1993.

Books

1. (a) "Introduction to Modern Liquid Chromatography," 2nd ed. L. R. Snyder and J. J. Kirkland. John Wiley & Son, New York, NY, 1979, 863 pp.

   (b) "Practical HPLC Method Development" L. R. Snyder, J. L. Glajch, and J. J. Kirkland. John Wiley & Sons, New York, NY, 1988.

2. "Reversed-Phase High Performance Liquid Chromatography." A. M. Krstulovic and P. R. Brown. John Wiley & Sons, New York, NY, 1982, 296 pp.

3. "Liquid Chromatography in Clinical Analysis." P. M. Kabra and L. J. Marton, editors. The Humana Press, Clifton, NJ, 1982, 466 pp.

4. "Basic Liquid Chromatography." E. L. Johnson and R. Stevenson. Varian Associates, Inc., Palo Alto, CA, 1978, 354 pp.

5. "Introduction to Mass Spectrometry." 2nd. ed. J. Throck Watson. Raven Press, New York, NY, 1985, 351 pp.

6. "Interpretation of Mass Spectra." 4th. ed. Fred W. McLafferty and František Tureček. University Science Books, Mill Valley, CA, 1993, 371 pp.

7. "Protein Purification." J. C. Jansen and L. Ryden, editors. VCH Publishers, New York, NY, 1989, 502 pp.

8. "Personal Computers for Scientists." Glenn I. Ouchi. American Chemical Society, Washington, DC, 1987, 276 pp.

Index